INORGANIC MEMBRANES
SYNTHESIS, CHARACTERISTICS
AND APPLICATIONS

INORGANIC MEMBRANES
SYNTHESIS, CHARACTERISTICS
AND APPLICATIONS

By

Ramesh R. Bhave, Ph.D

VNR VAN NOSTRAND REINHOLD
_____ New York

· 4452161

CHEMISTRY

Published by Van Nostrand Reinhold
115 Fifth Avenue
New York, New York 10003

Chapman and Hall
2–6 Boundary Row
London, SE1 8HN, England

Thomas Nelson Australia
102 Dodds Street
South Melbourne 3205
Victoria, Australia

Nelson Canada
1120 Birchmount Road
Scarborough, Ontario M1K 5G4, Canada

16 15 14 13 12 11 10 9 8 7 6 5 4 3 2 1

Library of Congress Cataloging-in-Publication Data

Inorganic membranes: synthesis, characteristics, and applications/
    edited by Ramesh R. Bhave.
        p.   cm.
    Includes bibliographical references and index.
    ISBN 0–442–31876–6
        1. Membranes (Technology)   I. Bhave, Ramesh R., 1952–.
TP159.M4I56   1991   91–18774
660'.2842—dc20   CIP

Contributors*

Prof. Anthony J. Burggraaf, Ph.D, The University of Twente, Laboratory for Inorganic Chemistry, Material Science and Catalysis, Enschede, The Netherlands

Prof. B. Tarodo de la Fuente, Ph.D, University of Montpellier, Sciences et Techniques du Languedoc, Montpellier, France

Jacques Gillot, Societe des Ceramiques Techniques, (a subsidiary of Alcoa Separations Technology, Inc.) Tarbes, France

Jacques Guibaud, Societe des Ceramiques Techniques, (a subsidiary of Alcoa Separations Technology, Inc.) Tarbes, France

H. Phillip Hsieh, Ph.D, Alcoa Laboratories, Alcoa Center, Pennsylvania

Klaas Keizer, Ph.D, The University of Twente, Laboratory for Inorganic Chemistry Material Science and Catalysis, Enschede, The Netherlands

Prof. Michel Rumeau, University of Montpellier, Laboratoire Genie des Procedes Montpellier, France

Robert J. Uhlhorn, Ph.D, The University of Twente, Laboratory for Inorganic Chemistry, Material Science and Catalysis, Enschede, The Netherlands (currently with UNILEVER, Detergents Division, Netherlands)

Venkat K. Venkataraman, Ph.D, U.S. Department of Energy, Morgantown Energy Technology Center, Morgantown, West Virginia, (*formerly* with Norton Company and Millipore)

Vassilis T. Zaspalis, Ph.D, The University of Twente, Laboratory for Inorganic Chemistry, Material Science and Catalysis, Enschede, The Netherlands

* (please note that contributors affiliation does not indicate that the work is supported by the contributors organization)

Foreword

Conventional separation techniques such as distillation, crystallization, filtration or solvent extraction were enriched in the 1960s by another class of processes which uses membranes as the principal separation elements.

The term membrane covers a great variety of materials, structures and geometries; the membrane acts as a selective barrier; under the effect of a driving force, on contact with chemical mixtures, it allows the separation of the constituents as a function of their specific transfer properties and is thus described as permselective.

Membrane separation techniques are very attractive and, in many cases, faster, more efficient and more economical than conventional processes, as the fractionation takes place without a phase change. Their use is expanding rapidly and the average annual growth rate over the next decade is estimated to be 10%.

Inorganic membranes for separations in liquid media appeared on the market in the beginning of the 1980s. These were porous, permeable, ceramic-type membranes with a composite structure and were called 3rd generation membranes to distinguish them from the purely organic membranes, which were the only ones used until then in separation applications.

The launching of this new generation of membranes was relatively slow; in the symposia and conferences on membrane separation techniques they occupied a relatively marginal place. It was only in 1989 that the First International Conference on Inorganic Membranes (ICIM) was organized at Montpellier (France). However, no basic text on the science and technology of inorganic membranes has yet been published.

This book by Dr. Ramesh R. Bhave will very well fill this gap. In the different chapters contributions by specialists, who are recognized authorities in the international research or industrial community, are included to treat all aspects relating to inorganic membranes very completely and objectively: from their history to fundamental phenomena, from methods of synthesis to characterization techniques and from present applications to medium- and long-term perspectives in separation processes or chemical reactions.

Research workers and engineers will find precise answers to their requirements and will also be able to thoroughly study specific points by using the extensive list of references at the end of each chapter.

Historically, it was in 1958, in the Second International Conference on the Peaceful Uses of Atomic Energy at Geneva, that the first papers on the production and the characterization of inorganic membranes were presented and discussed.

The French Atomic Energy Commission was particularly interested in the production of different kinds of microporous materials, metallic or ceramic, obtained by various techniques, from powder technology to the anodic oxidation of aluminum or chemical treatment of alloys. It recommended, in addition, the concept of a composite membrane, which was the only means of combining high permeability with mechanical strength.

This work was undertaken as part of the program for the separation of uranium isotopes by gaseous diffusion of uranium hexafluoride vapor. Paradoxically, it showed that the requirements for the porous texture of membranes used in ultrafiltration and microfiltration for separation in liquid media were similar to those of the components of gaseous diffusion barriers.

With the cooperation of industrial ceramic firms, the first uranium enrichment plant was put into operation in 1966 and the European Eurodif plant was operational in 1978. This plant, with more than 120 million barriers, the equivalent of 4 million square meters of installed surface, is believed to be the largest in the world equipped with inorganic membranes.

It is not, therefore, entirely fortuitous that the principal industrial firms that produce different types of inorganic membranes for ultrafiltration and microfiltration were involved in the French uranium enrichment program.

The development of ultrafiltration and microfiltration separation techniques using inorganic membranes had a laborious beginning. Gaining a foothold in a market dominated by organic membranes was difficult. It was necessary to convince industry that the additional cost of the inorganic membranes was largely compensated by greater lifetime or higher reliability.

In many situations the inherent properties of inorganic membranes clearly made them the only potential solution. It was here that inorganic membranes took birth in diverse applications involving agro-food, biotechnology and pharmaceutical products. It is believed that the development of membranes with higher selectivities, the nanofiltration membranes, will contribute to extend their use in the above listed relatively established fields and to a greater use in the chemical and petroleum industries and in environmental applications.

It was also possible, through the modifications of membrane surface, (which altered the wetting characteristics and surface charge), to produce an organic–inorganic membrane. These membranes may offer excellent mechanical and thermal resistance with the additional advantage of greater flexibility due to functionalization of organic radicals.

One of the essential improvements is the control of the interactions between the solution deposited on the pore walls to limit plugging and thus to increase the flux through the membrane. Another unique feature of charged organic–inorganic ultrafiltration membranes is their ability to combine one or more separation mechanisms. These membranes may also find use in environmental applications.

A major advantage of inorganic membranes lies in the possible applications of the numerous methods for studying porous membranes and separation barriers which have been perfected in France and the United States. These methods, in particular, have enabled the characterization of their intrinsic properties accurately and reproducibly.

This characterization, used profitably in determining the important production parameters, can ensure excellent control of the fabrication process and also allow the quantitative determination of possible anomalies under plant operating conditions. The ability to characterize inorganic membranes have enabled membranologists to obtain solutions to many problems by a less empirical approach.

Inorganic membranes also show significant potential for use in fields other than separation in liquid media. In the fractionation of gases or vapors, outside of isotopic separation, inorganic membranes allow us to envisage applications thanks to the advances in materials science and knowledge of surface phenomena and transport phenomena. In some situations, the selectivity is greatly influenced by the differences in surface fluxes or the preferential capillary condensation of the species being separated.

Recently, another method for the separation of gases, frequently described as "facilitated" transport membranes, using molten salts immobilized in inorganic (e.g. ceramic) porous supports has been studied.

A particularly innovative application of inorganic membranes is their use as a conversion catalyst in chemical syntheses, with or without separation of the species formed.

Inorganic membranes have established their use in a number of commercial applications involving agro-food, biotechnology and pharmaceutical products. However, the present development efforts to produce high-selectivity nanofiltration membranes, the perspectives for new chemical and environmental industry applications and the potentialities in the separation of gases or in chemical syntheses will hasten the deployment of inorganic membranes in separations. The annual growth rate for the use of inorganic membranes alone, is estimated to be in the range 15–20% over the next decade.

To fully realize its potentialities, the science of inorganic membranes will require contributions from many different fields: materials science, hydrodynamics, physical chemistry of surfaces and chemical engineering, thus necessitating the cooperation of groups of specialists.

There is no doubt that this first comprehensive volume describing the synthesis, characterization and applications will make their advantages and possibilities better known, will contribute to their development and will give rise to new collaborations. It will also be very useful to future users of inorganic membranes and will allow them to include membrane techniques in the research and optimization of their processes, with the assurance of respect for the environment.

Jean Charpin, Ph.D   Prof. Louis Cot
CEA, France   University of Montpellier,
                                            France

Preface

Inorganic membranes, since their introduction to commercial applications in the early 1970s, are making rapid inroads in many areas such as food and beverage processing, biotechnology applications and water treatment. There are also a couple of emerging areas such as the use of inorganic membranes in high-temperature gas separations and catalytic reactors.

As the use of inorganic membranes has expanded, a large number of publications covering a variety of subjects have appeared in the literature. This project was undertaken to fulfill the need to organize and present the wealth of information in a single reference volume describing the state-of-the-art in the development of inorganic membrane technology.

The book can be used by any scientifically trained individual interested in the field of inorganic membranes. The simple treatment of all subject matter will help the inexperienced user to easily comprehend the technical information presented in the book. The book will also serve as a useful reference volume for chemical engineers, chemists and technicians as well as business managers familiar with membrane separations.

I have been fortunate to be associated in the field of membrane separations for the past ten years. The past five years of my research and development efforts have been devoted to the emerging field of inorganic membranes. I have been truly intrigued by the great potential these materials show to solving difficult processing problems, including hostile environments where materials of high mechanical strength and structural integrity are required. Although inorganic membranes today enjoy only a small fraction ($\sim 5\%$) of the total membrane market, by the year 2000 their market share is expected to increase to about 10–12%.

In late 1987 when VNR first approached me to write the book, what inspired me most was the opportunity to describe and critique the developments in this continuously evolving field that holds great promise to displace conventional technologies with significant economical and technological advantage.

This book represents the first effort to document the available information in a logical and systematic fashion with in-depth technical discussion of fundamental principles and technical aspects related to the use of inorganic membranes. The subject matter is vividly illustrated through the use of numerous documented research and development studies describing specific membranes, material properties or applications. I hope that the book will serve to fulfill the needs of both current and prospective users in aspects related to the synthesis, characteristics (including characterization techniques) and applications.

The first chapter of the book is meant to provide the reader with a historical perspective on the developments in the field of inorganic membranes, which date back to the 1940s. The first application of inorganic membranes (and incidentally the world's largest) was for the enrichment of ^{235}U from an isotopic mixture of $^{235}UF_6$ and $^{238}UF_6$ by gaseous diffusion. The first commercial use of inorganic membranes was not realized until the early to mid-1970s for liquid separations using microfiltration or ultrafiltration membranes.

A large number of membrane synthesis techniques have been explored, developed and commercialized in the past 15–20 years. Chapter 2, on the synthesis of inorganic membranes, covers many aspects of these product research and development efforts in detail. The reader is provided with an overview of basic principles, theoretical aspects and membrane modification techniques. This may be helpful in the proper selection of membranes to suit specific application needs.

The general characteristics of inorganic membranes are reviewed in Chapter 3. Microstructural aspects such as pore diameter, thickness and membrane morphology are described along with consideration of material characteristics particularly those related to chemical, surface and mechanical properties.

The next two chapters (Chapters 4 and 5) are devoted to a thorough discussion of liquid permeation and separation, operating considerations and the design of filtration systems. During membrane filtration a variety of phenomena influence flux and/or separation behavior. These are identified and treated in some detail. A good understanding of fouling mechanisms can be very helpful in minimizing flux decline and in the proper selection of pore diameter and operating conditions.

Models for the prediction of microfiltration and ultrafiltration flux are described. Parameters influencing solute retention properties during ultrafiltration are identified and the subject is treated in some detail with illustrative examples.

In Chapter 5, cross-flow membrane filtration is described in detail to help the user understand the influence of important operating parameters on system performance. A few commonly used operating configurations are also described to point out their specific advantages or limitations. In addition, some aspects of system design and operation are discussed which can substantially influence system performance.

The current development status in the two emerging areas of gas separations and catalytic membrane reactors is reviewed in Chapters 6 and 7, respectively. Although inorganic gas separation membranes have been employed in nuclear (defense-related) applications, their use in commercial applications is yet to be realized. This is largely due to the lack of suitable commercial membranes with acceptable flux and/or separation performance under industrial conditions. The fundamentals of gas separations under various separation mechanisms are reviewed.

Another emerging area that is gaining considerable attention is the use of inorganic membrane reactors to enhance the productivity of chemical processes which are otherwise limited by thermodynamic constraints. The technical feasibility of high-temperature catalytic membrane reactors is being evaluated at the laboratory scale for a number of industrially important processes. Substantial development efforts are required to realize the full potential of inorganic membranes in commercial gas phase separations. The subject of liquid phase catalytic membrane reactors is not covered here due to the relatively few publications describing the use of inorganic membranes in these situations.

Key technical barriers are the production of small pore diameter (1–2 nm or smaller) porous composite membranes or high-flux dense membranes, and long-term high-temperature stability under industrial operating conditions. In addition, the design and fabrication of membrane elements, housings and seals will also pose significant technical challenges. Despite these difficulties, R & D efforts continue worldwide to develop improved membrane structures, since only inorganic membranes offer the possibility to withstand high-temperature processing environments. It is anticipated that a few of these applications will become commercially viable by the turn of the century.

Yet another emerging application is the use of composite ceramic membranes as gas filters (e.g. Membralox® alumina membranes) in the microelectronics industry. In this application, industrial gases are required to be purified to exceptionally high-purity levels. There are no published reports describing the performance of ceramic gas filters at this time. For this reason, applications of ceramic gas filters are not described in the volume. As this potentially rewarding area grows in the next few years their performance descriptions will most certainly appear in the scientific and technical literature.

The last two chapters of the book (Chapters 8 and 9) describe commercial and developing liquid phase applications (primarily filtration) that span many diversified industries. At the present time, almost all industrial plants using inorganic membranes are concerned with liquid phase separations. A large majority of these industrial filtration plants are in food, dairy and beverage processing. Biotechnology-related applications are still largely under development with a few exceptions such as cell harvesting and enzyme separations. Chapter 8 provides a complete review of all important commercial and developing food and biotechnology applications. Inorganic membrane performances in many other applications are not described in the open literature due to proprietary considerations.

The use of inorganic membranes for the filtration of water, wastewater treatment and chemical process industry filtration applications is covered in Chapter 9. The filtration of surface and groundwater to produce potable water is perhaps the most prominent industrial application. The treatment and/or recovery of oils from oily wastes and oil–water microemulsions is an area which has received a great deal of attention with some commercial success.

In the 1980s, a substantial number of inorganic membrane manufacturers emerged. A complete list of inorganic membrane manufacturers and their locations is given in the Appendix.

The contributions made in this volume by the author represent his personal and professional views and technical interpretations largely based on published work. These do not in any way reflect the views or endorsements by Alcoa Separations Technology, the author's current employer. This is also true of all contributors to this volume.

I want to express deep appreciation to Terry Dillman of Alcoa Separations for his personal interest and support. I also want to take this opportunity to thank A. J. Burggraaf, J. Gillot, J. Guibaud, H. P. Hsieh, K. Keizer, R. Rumeau, B. Tarodo de la Fuente, V. Venkataraman, R. Uhlhorn and V. Zaspalis for their extremely valuable contributions to this volume. Needless to say that without their cooperation and support it would not have been possible to present a comprehensive treatment of the developments in this emerging field, from membrane synthesis to gas and liquid phase applications.

I want to especially acknowledge the strong support and many sacrifices of my wife, Seema, during the writing of this book. Her understanding and appreciation made it possible to complete this project which consumed many many weekends, evenings and nights.

I also want to express my sincere appreciation to the VNR editorial staff and particularly to Marjorie Spencer and Alberta Gordon, for their patience, advice, and encouragement throughout the course of publication of this volume.

Contents

INORGANIC MEMBRANES
SYNTHESIS, CHARACTERISTICS
AND APPLICATIONS

1. The Developing Use of Inorganic Membranes: A Historical Perspective

J. GILLOT*

Societe des Ceramiques Techniques
(a subsidiary of Alcoa Separations Technology, Inc.) Tarbes

1.1. INTRODUCTION

To most users, inorganic membranes are a relatively new product. But in fact, their development started in the 1940s and can be schematically divided into three periods:

1. The development and mass production of membranes for the separation of uranium isotopes by the process of gaseous diffusion applied to UF_6.
2. Starting from this basis, the development and industrial use of a new generation of membranes adapted to the ultrafiltration and microfiltration of process liquid streams.
3. The more recent research work on a much broader range of membrane types aiming at separations using a variety of basic processes, including the coupling of catalytic reactions and membrane separation.

Many aspects of the development of uranium enrichment membranes were, and to a large extent still are classified, the scanty traces in the public domain only being a number of patents (CEA 1958, Clement, Grangeon and Kayser 1973, Miszenti and Mannetti 1971, Veyre et al. 1977). Much of the work done at present to develop new and improved inorganic membranes is also more or less classified.

Although the author participated in all three periods, this historical perspective covers only the first two periods. The perspective on the important developments in the third period as described above can be found throughout the book (especially in Chapters 6 and 7).

1.2. THE NUCLEAR PERIOD

Naturally occurring uranium contains a very small percentage (0.7%) of the fissile ^{235}U isotope to be used either in nuclear weapons, which require a very

* With R. R. Bhave.

1

high concentration, or in power generating plants, most of which require a concentration of approximately 3%. Since the separation by mass spectroscopy used in the Manhattan Project during World War II is prohibitively expensive, nuclear industries of the major industrialized countries researched to develop economically acceptable industrial processes.

Gaseous diffusion technology was thus developed and is still in use as the world's largest-scale industrial application using inorganic membranes. This process uses UF_6, which is the most practical or rather the least inconvenient volatile compound of uranium. The gas transport occurs by the Knudsen diffusion mechanism across a porous membrane with a pore diameter typically in the range 6–40 nm (Charpin and Rigny 1990). The lighter $^{235}UF_6$ molecule flows a little bit faster than the $^{238}UF_6$ molecule. The theoretical enrichment factor is 1.0043. In practice, the value is somewhat lower. This indicates that even for separating natural uranium into an enriched fraction containing approximately 3% ^{235}U and a depleted fraction containing approximately 0.2% ^{238}U, over 1000 stages will be required (e.g. there are 1400 stages in the Eurodif plant). Uranium enrichment plants are gigantic.

The qualitative problems involved in this development also were formidable. UF_6 is chemically very aggressive, which limits the choice of possible materials. Some metals and ceramics are among the candidate materials. Typically, tubular membranes were developed, which comprised a macroporous support, one or several intermediate layers of decreasing thickness and pore diameter, and the separating layer. The separating layer covered the internal surface of the tube (Charpin and Rigny 1990).

Little has been published on the work performed in the 1940s and 1950s. The first work was performed in the U.S.A. within the framework of the Manhattan Project in the 1940s (Egan 1989). In France, the Commissariat a l' Energie Atomique (CEA) began research on such membranes in the 1950s. At least three French industrial companies developed tubular macroporous supports for the CEA:

1. Desmarquest, a ceramics company which now is a subsidiary of Pechiney
2. Le Carbone Lorraine, a producer of carbon and graphite products (now also a subsidiary of Pechiney)
3. Compagnie Generale d' Electroceramique (CGEC), a ceramics company, then a subsidiary of Compagnie Generale d' Electricite (CGE), which later became Ceraver, the Membrane Department of which now belongs to Societe des Ceramiques Techniques (SCT), a subsidiary of Alcoa

Simultaneously, SFEC (Societe de Fabrication d' Elements Catalytiques) was created as a subsidiary of CEA to develop and manufacture the separating layer to be deposited on the macroporous support and to assemble the membrane into large modules.

In France, the first period of industrial production (late 1960s and early 1970s) was aimed at making the membranes for the Pierrelatte military enrichment plant. Most of the tubular membrane supports were made by CGEC, whereas the layers were made by SFEC. Since a good gaseous diffusion membrane does not wear out, it is very noteworthy that the original membranes are still in operation at this plant.

After the oil crisis in 1973, the need for large enrichment capacities for supply of fuel to the nuclear power plants became obvious and several European countries (Belgium, France, Italy and Spain) decided to build the huge Eurodif gas diffusion plant. This plant is located in France, in the Rhone valley, a few kilometers away from the Pierrelatte plant. Simultaneously, England, West Germany and the Netherlands (the Troika) chose to jointly develop the centrifugation process for uranium enrichment, which does not use membranes.

For Eurodif and for Pierrelatte, the supports were made by private industrial companies, the final separating layer by SFEC and the CEA developed the process and had the overall technical responsibility. A handful of companies were competing to manufacture the membrane support structure. Finally, two companies proposing ceramic oxide based supports, Ceraver (the new name of CGEC) and Euroceral (a 50/50 joint venture between Norton and Desmarquest) each won 50% of the market. This happened in 1975. Within a matter of 6 years, each company had to deliver more than 2,000,000 m^2 of supports which SFEC would convert into more than 4,000,000 m^2 of membranes (Charpin and Rigny 1990). Special plants were built at a very rapid pace. These were close to Tarbes for Ceraver, close to Montpellier for Euroceral and close to the Eurodif site for SFEC.

The enrichment capacity of Eurodif is 10,800,000 UTS (units of separation work). This corresponds to the fuel consumption of 90 nuclear reactors of the 900 MW class. In view of all the programs for building nuclear power plants hastily set up by many countries shortly after the 1973 oil crisis, it was clear that another uranium enrichment plant of similar size would have to be built immediately after Eurodif was completed. This was the Coredif project.

A few years later, one had to realize that most of the ambitious plans for building nuclear power plants would be strongly delayed or even abandoned. Only France stuck to its original plans and built a large number of nuclear reactors. The Coredif project did not materialize. In 1982, the membrane production plants of Ceraver, Euroceral and SFEC were shut down and later on dismantled due to the lack of demand for another enrichment plant (the service life of a gaseous diffusion membrane is several decades). For France, this was an abrupt end of the nuclear period for membranes.

In the U.S. similar developments in the area of inorganic membranes had taken place earlier on a somewhat larger scale and had resulted in membranes that are believed not to be of a ceramic nature (Charpin and Rigny 1990). Several very large gaseous diffusion plants were constructed (Oak Ridge and later Paducah in 1953 and Porsmouth in 1954) and are operated by the Department of Energy (DOE). These are not in operation but have at least twice the capacity of the Eurodif plant. At least part of the membrane development was made by Union Carbide, a company which had some impact on later developments.

In the USSR also, inorganic membranes were developed and gaseous diffusion plants were constructed to meet the needs for enriched uranium. For understandable reasons, very little is known of these developments.

In today's world, it is obvious that uranium enrichment by Knudsen diffusion has no future. Uranium enrichment by using laser technology can be accomplished much more efficiently. Such plants are expected to be ready for industrial-scale production by the year 2000 or 2005, when, according to current estimates, new uranium enrichment plants will be needed.

During the nuclear period, it was demonstrated that inorganic membranes, particularly the ceramic membranes, can be produced on a very large scale with exceptionally high quality to meet very stringent specifications. Further, the inorganic membranes were found to be very reliable and with long service life even under chemically aggressive environments. As indicated here, the membranes used in the Pierelatte plant are still performing very well even after more than 8 years of service. Likewise, at the time of this writing, the Eurodif membranes have satisfactorily completed 20 years of service and are expected to continue their performance for many years to come.

1.3. THE DEVELOPMENT OF ULTRAFILTRATION AND MICROFILTRATION INORGANIC MEMBRANES: THE 1980–1990 PERIOD

The development of industrial inorganic ultrafiltration (UF) and microfiltration (MF) membranes resulted from the combination of three factors:

1. the know-how accumulated by the companies that built the nuclear gaseous diffusion plants
2. the existence of ultrafiltration as an industrial process using polymeric membranes
3. the limitations of polymeric membranes in terms of temperature, pressure and durability

All major industrial participants in the developments that took place in the period 1980–1985, were companies which actively participated in the development and manufacture of inorganic membranes for nuclear applications, especially in the French nuclear program. The pioneering work was performed by two companies that were most active in this program, namely SFEC and Ceraver. The concept of inorganic ultrafiltration or microfiltration membranes is not new. The basic structure of these membranes is not different from that of the gas diffusion membranes described in a number of patents issued in the early 1970s (Clement, Grangeon and Kayser 1973, Miszenti and Mannetti 1971). The first attempt to use the high mechanical resistance of inorganic supports probably dates back to the 1960s, when dynamic membranes made of a mixture of zirconium hydroxide and polyacrylic acid deposited on a porous carbon or ceramic support were developed by the Oak Ridge National Laboratory in the U.S. (Kraus and Johnson 1966, Marcinkowsky, Johnson and Kraus 1968). These nonsintered or dynamically formed membranes require frequent regeneration by filtering through the support a suspension of zirconium hydroxide and polyacrylic acid. The thin cake thus formed is the separation layer. Such membranes later evolved into the ultrafiltration or reverse osmosis membranes made of a dynamic zirconium hydroxide on a stainless steel support that are now marketed in the U.S. by CARRE, a subsidiary of du Pont.

This concept later evolved into the Ucarsep® membrane made of a layer of nonsintered ceramic oxide (including ZrO_2) deposited on a porous carbon or ceramic support, which was patented by Union Carbide in 1973 (Trulson and Litz 1973). Apparently, the prospects for a significant industrial development of these membranes were at the time rather limited. In 1978, Union Carbide sold to SFEC the worldwide licence for these membranes, except for a number of applications in the textile industry in the U.S. At that time, SFEC recognized the potential of inorganic membranes, but declassification of the inorganic membrane technology it had itself developed for uranium enrichment was not possible.

The Union Carbide ultrafiltration membrane comprised a layer of unsintered ZrO_2 particles on a tubular carbon support with 6 mm inner diameter. Using the experience acquired in the nuclear program, SFEC added the step of sintering the ZrO_2 layer, thereby permanently attaching it to the support. SFEC also designed modules, filtration systems and developed ultrafiltration applications. In 1980, SFEC began selling complete ultrafiltration plants under the trademark of Carbosep®.

Since membranes no longer had important nuclear applications in future, SFEC was sold in 1987 by the CEA to the French company Rhone–Poulenc which merged them with their polymeric membrane division to form the new subsidiary, currently known as Tech Sep. ZrO_2-based ultrafiltration membranes on 6 mm inner-diameter carbon tubes continues to be the main product line of Tech Sep in terms of inorganic membranes.

Ceraver's entry into the microfiltration and ultrafiltration field followed a completely different approach. In 1980, it became apparent that the type of product made by Ceraver for uranium enrichment, which was a tubular support and an intermediate layer with a pore diameter in the microfiltration range, might be declassified. Ceraver therefore developed a range of α-Al$_2$O$_3$ microfiltration membranes on an α-Al$_2$O$_3$ support with two key features: first, the multichannel support and second, the possibility to backflush the filtrate in order to slow down fouling.

The use of a multichannel support made of a sintered oxide carrying a separation layer deposited on the surface of the channels was not a new concept. This was described in the patent literature as far back as the 1960s (Manjikian 1966). The multichannel geometry is particularly attractive in terms of its sturdiness, lower production cost compared to the single tube or tube-bundle geometry and lower energy requirement in the cross-flow re-circulation loop. However, Ceraver was the first company to industrially produce multichannel membranes. Since 1984 these membranes, which have 19 channels per element with a 4 mm channel diameter are sold under the trademark Membralox®.

The second innovative feature of the Membralox® membranes was the possibility to backflush, a feature the Carbosep® membranes did not offer. The principle of backflushing was not new, but it was the first time this was demonstrated with an industrial cross-flow membrane module. The ability to backflush can be very beneficial because in numerous applications fouling decreases the flux through a microfiltration membrane down to roughly similar range of values as those obtained with an ultrafiltration membrane. Backflushing is thus necessary to fully exploit the possibility of high flux offered by the relatively larger pore diameter (0.2 μm and larger) of micro-filtration layers.

SFEC was essentially able to market their ZrO$_2$-based ultrafiltration membranes to an already existing market in the sense that these membranes replaced polymeric UF membranes in a number of applications. They also developed a certain number of new applications. For Ceraver, the situation was different. When the Membralox® membranes were first developed, microfiltration was performed exclusively with dead-end polymeric cartridge filters. In parallel to the development of inorganic MF membranes, Ceraver initiated the development of cross-flow MF with backflushing as a new industrial process.

The first generation Membralox® membranes were essentially developed for MF applications. Although γ-Al$_2$O$_3$ (also described as transition Al$_2$O$_3$) UF layers with pore diameters suitable for UF were available, their poor chemical resistance prevented their widespread use for UF applications.

Membralox® UF membranes with ZrO_2 layers were commercialized in 1988. These are resistant to extreme pH values and can be backflushed.

Ceraver's business approach was, however, completely different from that of SFEC. Ceraver's strength was primarily in the manufacture of technical ceramics. Thus, Ceraver sold membranes in the form of complete modules to equipment manufacturers who developed the filtration systems including in most cases the filtration process itself.

In 1986, CGE, which by then had its primary focus in the energy and communication businesses, divested its association from materials, and sold the ceramic part of Ceraver, including the ceramic membranes division to Alcoa. Under the name SCT it is now a subsidiary of the recently formed Alcoa Separations Technology, Inc.

A few other players in the nuclear membranes activity also developed inorganic membranes for the filtration of liquids. This was the case with Norton-USA who with the know-how of Euroceral developed MF membranes made of an α-Al_2O_3 tubular support with an α-Al_2O_3 layer. The inner tube diameter was 3 mm and the outer diameter 5 mm. In 1988–1989, Norton also produced the multichannel membrane elements. These membranes produced by Norton are now sold by Millipore under the trademark Ceraflo®.

In the early 1980s, former employees of Euroceral founded a small company located near Montpellier in France known as Ceram-Filtre. The rather less well-known Ceram-Filtre membranes comprise a multichannel support with 19 channels of 4 mm diameter and a microfiltration membrane made of an oxide.

Another participant in the French nuclear program, Le Carbone-Lorraine, developed inorganic membranes by combining their know-how in the field of membranes with their expertise in carbon. They developed tubular UF and MF membranes using a tubular carbon support (inner diameter 6 mm, outer diameter 10 mm). The carbon support is made of carbon fibers coated with and bonded by CVD carbon, the separating layers also being made of carbon. These membranes have been marketed since 1988.

The membrane research and development activities of some university laboratories is also a fallout of the nuclear membrane program. The inorganic membrane work performed by the University of Montpellier originated in a cooperative effort with the neighboring Euroceral plant. This cooperative effort continued with Ceraver subsequent to the shutdown of the Euroceral plant.

A completely different type of inorganic membrane also has its origin in the nuclear industry: the asymmetric alumina membranes obtained by the anodic oxidation of an aluminum sheet were first developed for uranium enrichment in the 1950s in France (Charpin, Plurien and Mammejac 1958) and in Sweden (Martensson et al. 1958). Such membranes are now marketed under the trademark of Anopore® by Anotec, A British subsidiary of Alcan.

1.4. THE THIRD STAGE IN THE DEVELOPMENT OF INORGANIC MEMBRANES: FURTHER DEVELOPMENTS BY OTHER ORGANIZATIONS

In the second half of the 1980s, an increasing number of companies entered the field of inorganic membranes, the most significant ones being ceramic companies such as NGK of Japan which also developed a multichannel membrane element (19 channels, 3 mm diameter), Nippon Cement and Toto also from Japan and very recently Corning who also developed a multichannel membrane structure.

An increasingly large number of university and industrial laboratories have also begun exploring new techniques for producing and/or utilizing existing inorganic membranes, including metal membranes, to develop new applications. A variety of new separating layers are under development, including porous and nonporous glasses, layers doped with catalysts, etc. The characteristics of many of these are discussed in the later chapters.

1.5. CONCLUSIONS

In summary, the development of inorganic membranes was initially oriented towards uranium enrichment which is still by very far their most significant application. Some of the key participants involved in the nuclear programs further developed them into cross-flow filtration membranes. The recent years have seen the start of a much broader exploration of the manyfold potentialities of inorganic membranes, both in terms of materials and applications. Thus, a multifaceted new field of technology is emerging.

REFERENCES

CEA. 1958. Porous membranes with very fine porosity and their production process. French Patent 1,197,982.

Charpin, J., P. Plurien and S. Mammejac. 1958. Application of general methods of study of porous bodies to the determination of the characteristics of barriers. *Proc. 2nd United Nations Intl. Conf. Peaceful Uses of Atomic Energy*, 4: 380–87.

Charpin, J. and P. Rigny. 1990. Inorganic membranes for separative techniques: From uranium isotope separation to non-nuclear fields. *Proc. 1st Intl. Conf. Inorganic Membranes*, 3–6 July, 1–16, Montpellier.

Clement, R., A. Grangeon and J. C. Kayser. 1973. Process for preparing filtering elements with high permeability. French Patent 2,527,092.

Egan, B. Z. 1989. Using inorganic membranes to separate gases: R/D status review. Oak Ridge National Laboratory Report ORNL/TM-11345.

Kraus, K. A. and J. S. Johnson. 1966. Colloidal hydrous oxide hyperfiltration membrane. U.S. Patent 3,413,219.

Manjikian, S. 1966. Production of semipermeable membranes directly on the surfaces of permeable support bodies. U.S. Patent 3,544,358.

Marcinkowsky, A. E., J. S. Johnson and K. A. Kraus. 1968. Hyperfiltration method of removing organic solute from aqueous solution. U.S. Patent 3,537,988.

Martensson, M., K. E. Holmberg, C. Lofman and E. I. Eriksson. 1958. Some types of membranes for isotope separation by gaseous diffusion. Proc. 2nd United Nations Intl. Conf. Peaceful Uses of Atomic Energy. 4: 395–404., Geneva.

Miszenti, G. S. and C. A. Mannetti. 1971. Process for preparing porous composite membranes or barriers for gaseous diffusion systems. Italian Patent 27802A/71.

Trulson, O. C. and L. M. Litz. 1973. Ultrafiltration apparatus and process for the treatment of liquids. U.S. Patent 3,977,967.

Veyre, R., S. Richard, F. Pejot, A. Grangeon, J. Charpin, P. Plurien and B. Rasneur. 1977. Process for producing permeable mineral membranes. French Patent 2,550,953.

2. Synthesis of Inorganic Membranes

A. J. BURGGRAAF and K. KEIZER*

University of Twente, Faculty of Chemical Technology, Enschede

2.1. INTRODUCTION AND OVERVIEW

2.1.1. General Background: Membrane Types and Structures

The aim of this introductory section is twofold. In the first place, the large variety of different synthesis methods and techniques will be placed against the background of membrane types and structures. These will be briefly summarized with focus on their relation with synthesis aspects. This will justify a selection of *two* groups of methods: those which will be discussed in more detail (Sections 2.3–2.9) and those which will only be mentioned, but not treated extensively (Section 2.8). In the second place, a brief summary of the most important aspects of membranes relating to synthesis methods will be given. This will serve as a guideline in the more detailed treatment of particular synthesis methods. The field of inorganic membranes has attracted more attention in recent years and is now rapidly developing. This is reflected in the relatively large number of patents or patent applications in the last few years indicating a considerable, partly hidden, industrial activity. Furthermore, the number of reviews since 1987 have sharply increased, as is the attention inorganic membranes receive in an increasing number of symposia and conferences. Many papers have a preliminary character and most contributions have a strongly descriptive nature. A focus will be given on those fields and contributions where at least a certain coupling between synthesis and resulting microstructure has been shown.

A membrane can be described as a semipermeable barrier between two phases which prevents intimate contact. This barrier must be permselective which means that it restricts the movement of molecules in it in a very specific way. The barrier can be solid, liquid or gas. Permselectivity can be obtained by many mechanisms:

1. Size exclusion or molecular sieving
2. Differences in diffusion coefficients (bulk as well as surface)
3. Differences in electrical charge
4. Differences in solubility
5. Differences in adsorption and/or reactivity on (internal) surfaces

* With R. R. Bhave.

The flux of liquids or gases through the membrane is in most cases driven by a pressure gradient and sometimes by an electric field gradient. Membranes can be used for:

1. Separation of mixtures (liquids, gases or liquid–solid mixtures can be separated).
2. Manipulation of chemical reactions: shifting the equilibrium situation or manipulation of the conversion or selectivity of catalytic reactions are two possibilities.

The effectiveness of the membrane in a certain application depends on the detailed morphology and microstructure of the membrane system, in addition to the performance of the above mentioned physicochemical mechanisms. These are critically determined by the synthesis process and this is why details of the preparation procedures are so important. The most important and well developed of these procedures are treated in Sections 2.3–2.9.

Inorganic membranes can be categorized as shown in Table 2.1. The *dense* inorganic membranes consist of solid layers of metals (Pd, Ag, alloys) or (oxidic) solid electrolytes which allow diffusion of hydrogen (or oxygen). In the case of solid electrolytes transport of ions takes place. Another category of dense membranes consist of a porous support in which a "liquid" is

Table 2.1. Types of Inorganic Membranes

	Main Characteristics	Comments
Dense	Metal foil (Oxidic) solid electrolyte Liquid immobilized (LIM) Permanent	Solution/diffusion of atomic or ionic species
Dynamic	Nonpermanent*	Ion exchange in hydroxide layers on a support
Porous metal or nonmetallic Inorganic	Symmetric, asymmetric Supported, nonsupported Pore shape, morphology and size Chemical nature of pore surface	Permselective diffusion affected by the pore characteristics
Composite Modified	Two-phase particle mixture Pores in matrix (partially) filled with 2nd phase sandwich structures	Distribution, important, microparticles, props

* "Nonpermanent" means separation layer is formed during the preparation process in-situ on a porous support

immobilized. The liquid fills the pores completely and is semipermeable. Interesting examples are molten salts immobilized in porous steel or ceramic supports and semipermeable for oxygen. Sometimes this liquid can be formed in-situ during the process under consideration. This is, for example, the case during the use of ceramic membranes in the decomposition of H_2S where liquid S is condensed in the pores and blocks hydrogen diffusion. Another example is the group of the so-called dynamic membranes where a hydroxide (gel) is precipitated in or on a porous support.

The *porous* membranes consist of a porous metal or ceramic support with porous top layers which can have different morphologies and microstructures. Their essential structural features are presented in Figures 2.1 and 2.2 and are discussed later (Section 2.2).

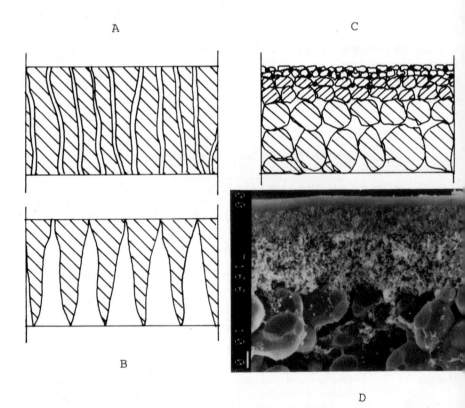

Figure 2.1. Schematic representation of main types of pore structures and membranes. A and B: homogeneous unsupported; straight pores C: supported asymmetric, interconnected pores D: a photograph of a membrane of the type (c). (SCT-support + γ-Al_2O_3 top layer UT Twente)

From Figure 2.1 it can be seen that there are three types of pore systems: straight pores running from one side of the membrane to the other with a constant pore diameter (Figure 2.1a) or conical pores (Figure 2.1b). These types of systems can be correlated with track-etch and anodic oxidation processes, respectively. The system shown in Figure 2.1c consists of a percolation system of pores with more or less regular shapes or with a spongy structure. This is correlated with packing of particles and phase separation, respectively. In composite and modified membranes the top layer (with very fine pores) shown in Figure 2.2 is modified further as schematically represented in Figure 2.3 or consists of an intimate mixture of two phases. The modification technologies in most cases consist of precipitation of a phase from liquids or gases followed by further treatments (see for further details Section 2.9) and result in a decrease of the pore size or in a change in the chemical character of the internal pore surface. The obtainable pore sizes are schematically represented in Figure 2.4 together with the most important related fields of application. Principles to obtain the required mean pore size and narrow pore size distribution are summarized below.

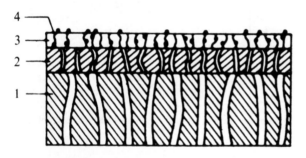

1. porous support (1–15 μm pores)
2. intermediate layer(s) (100–1500 nm)
3. separation (top) layer (3–100 nm)
4. modification of separation layer

1 + 2	is microfiltration
	or 'primary' membrane
1 + 2 + 3	is ultrafiltration
	or 'secondary' membrane
1 + 2 + 3 + 4	is hyperfiltration or/and
	gas separation membrane

Figure 2.2. Schematic representation of an asymmetric-composite membrane (Keizer and Burg-graaf 1988).

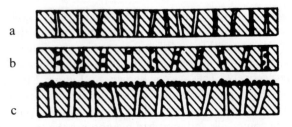

a   Homogenous (multi) layers in the pores
b   Plugs in the pores (constrictions)
c   Plugs/layers on top of the pores

Figure 2.3. Schematic representation of the microstructure of modified membrane top layers (Keizer and Burggraaf 1988).

Application fields

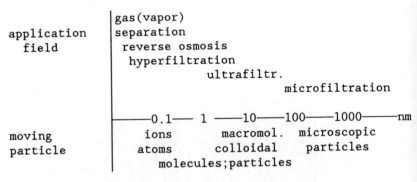

Figure 2.4. Pore size range of ceramic membranes and related application fields.

2.2. BASIC PRINCIPLES OF MEMBRANE SYNTHESIS

In this section a short introduction will be given on the synthesis of porous ceramic membranes by sol–gel techniques and anodization, carbon membranes, glass membranes and track-etch membranes. An extensive discussion will be given in Sections 2.3–2.8.

Terms such as symmetric and asymmetric, as well as microporous, mesoporous and macroporous materials will be introduced. Symmetric membranes are systems with a homogeneous structure throughout the membrane. Examples can be found in capillary glass membranes or anodized alumina membranes. Asymmetric membranes have a gradual change in structure throughout the membrane. In most cases these are composite membranes consisting of several layers with a gradual decrease in pore size to the feed side of the membrane. Examples are ceramic aluminas synthesized by the sol–gel technique or carbon/zirconia membranes.

Pore diameters larger than 50 nm are called macropores, mesopores have a diameter between 2 and 50 nm and below a diameter of 2 nm the system is called microporous (Sing et al. 1985).

2.2.1. Ceramic Asymmetric Membranes

The asymmetric membrane system shown in Figure 2.2 consists of a porous support a few millimeters in thickness, with pores in the range 1–10 μm, a porous intermediate layer of 10–100 μm thickness, with pores of 50–500 nm, and a top layer (the proper separation layer, e.g. for ultrafiltration) with a thickness of 1 μm (or smaller)–10 μm with pores of 2–50 nm. The intermediate layer must prevent the penetration of the precursor of the top layer material into the pores of the support during the synthesis and the collapse of the thin finished top layer into the large pores of the support. Furthermore, it helps to regulate the pressure drop across the top layer of the membrane in operation. In a number of cases it can be dispensed with (e.g. in carbon-supported zirconia membranes manufactured by SFEC). As shown in Figure 2.3 the pore system can be further modified in different ways. In all cases the top layer must be defect-free (no cracks or pinholes) and have preferably a narrow pore size distribution. This sets severe demands on the quality of the intermediate layer and of the support. It may also require development of special technologies to overcome inferior qualities of the support system. By far the most frequently used principle to meet these requirements is the formation of a layer consisting of a packing of well-ordered, uniform-sized particles. The size and shape of the particles determine the minimum obtainable mean size and pore size. These parameters as well as the porosity can be changed by further heat treatment. The main process for making ceramic membranes is to first prepare a dispersion of fine particles (called slip) and then to deposit the particles contained in the slip on a porous support by a slip-casting method. The capillary pressure drop created on letting the slip come into contact with the microporous support forces the dispersion medium of the slip to flow into the pores of the support. The slip particles are concentrated at the pore entrance to form a layer of particles or a gel layer. When relatively large particles are used to make the support or intermediate layer (with pore sizes > 1 μm) the particles can be precipitated from super-saturated solutions. These need to be calcined and classified by sedimentation, centrifuging or sieving. A slurry of these particles can be used after stabilization.

For making the top layers with very small pores, colloidal suspensions are used. Here we need nanometer-sized particles which are stabilized in a liquid dispersion medium by colloidal or other physicochemical methods. These particles and colloidal suspensions are obtained mainly by the so-called sol–gel methods. The essential feature of this technique is the controlled hydrolysis of an organometallic compound or salt and its subsequent peptization. Very critical is the drying step in the process because here large (capillary) forces can cause cracking of the layer very easily. This can be controlled by particle shape, agglomeration control, addition of binders, roughness of the support, etc. The above-mentioned method will be discussed further in Section 2.3. A summary of presently obtainable combinations of supports and unmodified top layers is given in Table 2.2. The minimum size of pores obtainable in this way is about 2.5 nm. This number is related to the smallest sol primary particle which can be obtained. The size of this primary particle is determined by seeding and crystallization parameters such as the surface tension (τ_s) and the free-enthalpy difference (ΔG) between dissolved and crystallized material. Stable nuclei smaller than 5 nm are hardly obtained. Thermal treatments always increase the pore size due to a decrease in the free-enthalpy G.

The pore shape is determined by the particle shape. Plate-shaped particles lead to plate-shaped pores in the case of regular packing. Sphere-shaped particles favor cylindrical or sometimes ink-bottle-type pores.

Table 2.2. Asymmetric Composite Membranes: Combinations of Substrates and Top Layers

Top Layer	Al_2O_3 1	ZrO_2 2	TiO_2 3	SiO_2 4	C 5	SiC 6	Binary Layers	Modified Layers
Substrate								
Alumina	Ind.*	Ind.	Lab.†	Lab.			1 + 2/1 + 3	Ag, MgO V_2O_5, SiO_2
Zirconia		Lab.						
Titania			Lab.					Ag, V_2O_5
Silica								
Carbon	Lab.	Ind.			Lab.			
SiC						Lab.		
Sintered:								
Steel			Lab.					
Nickel			Lab.					

* Ind.: Industry
† Lab.: Laboratory

To reduce further the pore size and/or to introduce specific interactions between the solid surface and the liquid or gaseous medium in the pores, sol–gel layers need to be modified. In principle this is done by precipitation or by adsorption of components from a gaseous or liquid medium followed by heat treatment of the formed products inside the pores or the pore entrance. This will be further discussed in Section 2.7.

2.2.2. Glass Membranes

Glass membranes with an isotropic spongy structure of interconnected pores can be prepared by thermally demixing a homogeneous $Na_2O-B_2O_3-SiO_2$ glass phase in two phases. The $Na_2O-B_2O_3$-rich phase is then acid-leached thereby creating a microporous SiO_2-rich phase (Hsieh 1988). Some porous metal membranes have been made in a similar way with a strong acid or other types of leachant (Hsieh 1988). The remaining silica glass structure is not very chemically resistant. This has been overcome partially by surface treatment (e.g. by means of chemical agents) of the internal pore structure which make the surface hydrophobic (Schnabel and Vaulont 1978). The advantage of glass membranes is that capillaries (hollow fibers) can be easily formed, and can be further modified as described above to porous hollow-fiber membranes.

2.2.3. Anodic Membranes

Pores with a linear form as shown in Figure 2.1 are produced by the so-called anodic oxidation process (Smith 1974). Here one side of a thin high-purity aluminum foil is anodically oxidized in an acid electrolyte. A regular pattern of pores is formed. The pore size is determined by the voltage used and by the type of acid, the pore shape being always conical. The process must be stopped before the foil is oxidized completely and to avoid closure of pores. The unaffected part of the metal foil is subsequently etched away with a strong acid. The resultant structure has distinctive conical pores perpendicular to the macroscopic surface of the membrane. The membranes so obtained are not stable under long exposure to water. The stability can be improved by treatment in hot water or in a base. Such a treatment can also be used to decrease the pore size on one side of the membrane, and as a consequence an asymmetric membrane can be produced. The disadvantage of the method is that only unsupported membranes can be produced in the form of membrane foils. To get sufficient mechanical stability they must be supported in some way for most applications.

2.2.4. Track-Etch Membranes

Pores with a very regular, linear shape can be produced by the track-etch method (Quinn et al. 1972). Here a thin layer of a material is bombarded with highly energetic particles from a radioactive source. The track left behind in the material is much more sensitive to an etchant in the direction of the track axis than perpendicular to it. So etching the material results in straight pores of uniform shape and size with pore diameters ranging between 6 nm and 1200 nm. To avoid overlap of pores only 2–5% of the surface can be occupied by the pores. This process has been applied on polymers (e.g. Nuclepore membranes) and on some inorganic systems like mica. Membranes so obtained are attractive as model systems for fundamental studies.

2.2.5. Pyrolysis

Membranes with extremely small pores (< 2.5 nm diameter) can be made by pyrolysis of polymeric precursors or by modification methods listed above. Molecular sieve carbon or silica membranes with pore diameters of 1 nm have been made by controlled pyrolysis of certain thermoset polymers (e.g. Koresh, Jacob and Soffer 1983) or silicone rubbers (Lee and Khang 1986), respectively. There is, however, very little information in the published literature. Molecular sieve dimensions can also be obtained by modifying the pore system of an already formed membrane structure. It has been claimed that zeolitic membranes can be prepared by reaction of alumina membranes with silica and alkali followed by hydrothermal treatment (Suzuki 1987). Very small pores are also obtained by hydrolysis of organometallic silicium compounds in alumina membranes followed by heat treatment (Uhlhorn, Keizer and Burggraaf 1989). Finally, oxides or metals can be precipitated or adsorbed from solutions or by gas phase deposition within the pores of an already formed membrane to modify the chemical nature of the membrane or to decrease the effective pore size. In the last case a high concentration of the precipitated material in the pore system is necessary. The above-mentioned methods have been reported very recently (1987–1989) and the results are not yet substantiated very well.

2.2.6. Dense Membranes

A second class of membranes are described as dense membranes. They may consist of thin plates of metals (Pd and its alloys, Ag and some alloys) or oxides (stabilized zirconia or bismuth oxides, cerates). These membranes are permeable to atomic (for metals) or ionic (for oxides) forms of hydrogen or oxygen and have been studied, especially, in conjunction with chemical reactions like (oxidative) dehydrogenation, partial oxidation etc. in membrane reactors. Their main drawback is the low permeability. This might be improved by making very thin layers (micrometer to nanometer range), e.g. by deposition in a pore system. A second form of "dense" membranes are the so-called liquid-immobilized membranes (LIM). Here the pores of a membrane are completely filled with a liquid which is permselective for certain compounds. In the polymer field this principle has been already investigated extensively. In the inorganic membrane field research efforts have just begun. With molten salts incorporated in a porous matrix, one can obtain permeabilities for oxygen or ammonia comparable with those of porous materials (Pez 1986, Dunbobbin and Brown 1987). Important parameters are the wettability of the matrix by the liquid and the morphology of the pore system because these determine the degree to which the liquid is captured (immobilized) within the membrane system.

2.3. PACKING OF PARTICLES FROM SUSPENSIONS

2.3.1. Introduction and Support Systems

Membranes produced by packing of particles from dispersions have the general structure as discussed in Section 2.2 and given in Figure 2.2. The thin top layer with (very) small pores is applied on top of a support system which consists of one or two much thicker layers with (much) larger pores. The main support consists of a packing of rather coarse-grained material (micron range) which is produced in a classical way by cold isostatic pressing of a dry powder, by co-extrusion of a paste of ceramic powder with additions of binders and plasticizers or by slip-casting (Messing, Fuller and Housner 1988). After burning away the organic material the so-called "green" compact is sintered. In order to obtain defect-free membranes, thin top layers on the support system must fulfill more stringent requirements than those utilized in the manufacture of commercially available porous tube materials. Pore size distribution and roughness must be smaller than usual. The ways to obtain these characteristics are largely classified with practically no published information. Some aspects have been discussed by Vuren et al. (1987) and Terpstra, Bonekamp and Veringa (1988). It is important to obtain a narrow pore size distribution. Therefore, suspensions or pastes are prepared from a powder having a narrow particle size distribution. This implies a very good control of the agglomeration state of the material by deagglomeration treatments (e.g. milling, ultrasonification) and/or removal of the fraction with the largest diameters e.g. by sedimentation. Organic surfactants are sometimes added to counteract flocculation of the deagglomerated suspension.

The quality of the support is especially critical if the formation of the top layer is mainly determined by capillary action on the support (see Section 2.3.2). Then, besides a narrow pore size distribution the wettability of the support system plays a role (see Equation 2.1). An example of the synthesis of a two-layer support and ultrafiltration membrane is given in the French Patent 2,463,636 (Auriol and Tritten 1973). In many cases an intermediate layer, whose pore sizes and thickness lie between those of the main support and the top layer (see Figure 2.2), is used. This intermediate layer can be used to improve the quality of the support system. If large capillary pressures are used to form such an intermediate layer, defects (pinholes) in the support will be "transferred" to this layer. This can be avoided by decreasing the acting capillary pressures or even by eliminating them. This can be done in several ways.

If a film coating technique is used, the viscosity of the system is increased to such an extent that none or hardly any penetration of the "liquid" in the pore system occurs. The film thickness obtained is governed by the surface tension and the viscosity of the suspension (Deryaquin and Levi 1964). The same result is obtained with co-extrusion of material of the main support and of the intermediate layer. The capillary driving force for extracting liquid from the coating suspension into the pores of the support can be decreased by filling the pores of the support material with a liquid, with a small difference in surface tension with respect to that of the suspension from which the intermediate layer (or top layer) is formed (Tallmadge and Gutfinger 1967).

In another method the pores of the support are rendered nonwettable for the coating suspension liquid. This can be obtained by forming a hydrophobic pore surface on the support in the case of an aqueous coating suspension. The support surface is treated, e.g. by organic silanes which react with surface hydroxyl groups as described by Messing (1978). Such a surface has at the same time improved properties for separation of a mixture of polar or nonpolar gases or hydrophobic and hydrophilic liquids. The disadvantage of a decreased capillary action in the formation of a layer is the diminished adherence of the support and that layer. The adherence as well as elimination of defect formation can be improved by a smaller roughness of the support material. Gillot (1987) reports a roughness which should be less than 10% of the mean particle diameter of the grain size of the support material. This implies again that the particle size distribution of the support should be narrow to obtain a locally well-defined roughness. Gillot uses this principle to make an improved three-layer membrane system, with particles of 0.55 μm in the top layer and using polyvinylalcohol to control the viscosity and a surfactant Darvan C to avoid flocculation. The mean pore size of the resulting top layer is 0.26 μm and a small roughness is reported. To obtain smaller pore sizes as indicated above it is necessary to use (ultra) fine-grained powders and suspensions in the synthesis of the top layer. This is obvious because in well-packed systems of uniform particles the mean pore radius is minimum (about 0.4–0.7 of the mean particle radius), depending on the packing structure. This means that (colloidal) suspensions with particle diameters ranging from 5 nm–100 nm are needed. These can be obtained in different ways described in recent symposia and congress proceedings (Hench and Ulrich 1984, 1986, Brinker, Clark and Ulrich 1984, 1988). The most commonly used route for membrane synthesis with packing methods is the *slip-casting* process using suspensions obtained by sol–gel processes. In the remainder of this chapter attention will be focussed on this combination.

2.3.2. Sol–Gel Process

The sol–gel process can be divided into two main routes which are schematically shown in Figure 2.5. These may be distinguished as the colloidal suspension route and the polymeric gel route. In both cases a precursor is hydrolyzed while simultaneously a condensation or polymerization reaction occurs. The essential parameter to control is the hydrolysis rate with respect to the polycondensation rate. The precursor is either an inorganic salt or a metal organic compound. The chemistry of the initial stages has been described by Livage (1986). In the *colloidal route* a faster hydrolysis rate is obtained by using a precursor with a fast hydrolysis rate and by reacting the precursor with excess water. A precipitate of gelatinous hydroxide or hydrated oxide particles is formed which is peptized in a subsequent step to a stable colloidal suspension. The elementary particle size ranges, depending on the system and processing conditions, from 3–15 nm and these particles form loosely bound agglomerates with sizes ranging from 5–1000 nm. The size of the agglomerates can be decreased, e.g. by ultrasonification of the suspension and by manipulation of the electrical charge on the particles. By increasing the concentration of the suspension and/or by manipulation of the surface (zeta) potential of the sol particles the colloidal suspension is transformed to a gel structure consisting of interlinked chains of particles or agglomerates (Figure 2.5). As discussed by Partlow and Yoldas (1981) the packing density at the time of gelation (i.e. the gelling volume) can vary from a rather loose to a dense form depending on the charge on the particles. This means that the pH and the nature of the electrolyte (or anion in the peptizing acid) has an important effect on the gelling point and volume, because they determine the mutual repulsion force which is necessary to obtain a stable colloidal suspension. The anion chosen for the electrolyte or peptizing acid must not form a complex with the metal ion of the membrane to be formed. In instances where the initial particle has a charge opposite to that of the electrolyte, the gelling volume exhibits a minimum with increasing electrolyte

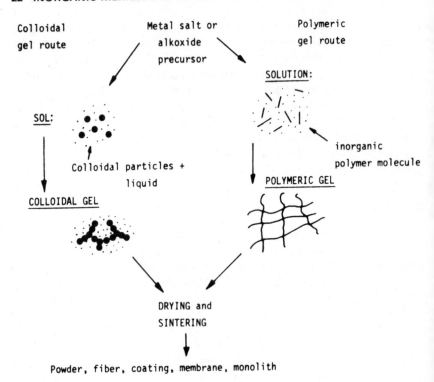

Figure 2.5. Scheme of sol–gel routes. Colloidal sol–gel route and polymeric gel route (Burggraaf, Keizer and van Hassel (1989a, b).

concentration. This is, for example, the case for alumina sols. It seems that gels with their minimum volume are better suited to obtain monolithic structures (Partlow and Yoldas 1981, Yoldas 1975) and membranes (Leenaars, Keizer and Burggraaf 1984, 1987, Leenaars and Burggraaf 1985). This means that pH, counter-ion type and concentration must be chosen in such a way that the particle is just far enough from its point of zero charge, i.e. isoelectric point (IEP), to prevent flocculation. With conditions too close to the IEP a poorly densified film will be obtained. In this way ordered aggregates of elementary particles can be obtained. In the further processing this is the best starting situation to give defect-free membranes (or monoliths) with a narrow pore size distribution.

The hydrolysis and polymerization rate of metal organic compounds can generally be better controlled than those of metal salts. The chemical reaction involves two steps (Livage 1986):

1. The partial hydrolysis of the metal organic compound (e.g. a metal alkoxide) introduces the active functional OH groups, attached to metal atoms.

2. These then react with each other or with other reactants to form a polymeric solution which further polymerizes to form a viscous solution of organic–inorganic polymeric molecules.

In the polymeric gel route the hydrolysis rate is kept low by adding successively small amounts of water and by choosing a precursor which hydrolyzes relatively slowly. The final stage of this process is a strongly interlinked gel network (Figure 2.5) with a structure different from that obtained from the colloidal route. This can be seen from the fact that the network formation takes place continuously within the liquid. The gel will form and shrink even within the liquid. It is not necessary to remove this liquid to obtain a gel as in the colloidal route. This means that concentrations of solid material in polymeric gels are usually smaller than in particle gels. The water necessary for the reaction can be supplied in different ways: (1) slowly adding a water or water/alcohol solution to an alcoholic solution of the alkoxide, (2) in-situ production of H_2O through an esterification reaction by adding an organic acid to the alkoxide solution, (3) dissolving an alkaline base or (4) an hydrated salt into the alkoxide solution in alcohol. The local water concentration can be manipulated in this way thereby strongly influencing the gel volume at the gelling point. With method (4) even complex compounds ($BaTiO_3$) can be obtained (Livage 1986). Finally, the gelation process can be significantly changed by the nature of catalysis of the polycondensation/polymerization reaction (Iler 1979). A model which predicts some aspects of the inorganic polymerization reactions is given by Livage and Henry (1985). Silica systems can be controlled very well and both colloidal and polymeric gel routes can be realized. Alumina has a strong preference to follow the colloidal route, while titania systems behave intermediately.

2.3.3. The Slip-Casting of Ceramic Membranes

A common method to slip-cast ceramic membranes is to start with a colloidal suspension or polymeric solution as described in the previous section. This is called a "slip". The porous support system is dipped in the slip and the dispersion medium (in most cases water or alcohol–water mixtures) is forced into the pores of the support by a pressure drop (ΔP_c) created by capillary action of the microporous support. At the interface the solid particles are retained and concentrated at the entrance of pores to form a gel layer as in the case of sol–gel processes. It is important that formation of the gel layer starts immediately and that the solid particles do not penetrate the pores of the support system. This means that the solid concentration in the slip must not be too low, the slip must be close to its gelling state, the particle (or agglomerate) size must not be too small compared with the pore size of the support, unless agglomerates are formed in the pore entrance immediately at the start of the process. Some variables such as solid concentration and particle diameter have been given by Cot, Guizard and Larbot (1988) who have demonstrated that the characteristics of the membrane and its formation can be influenced by the agglomerate state of the slip.

The smaller and more uniform the primary particles, and the weaker the agglomerates in the sol are, the smaller the pore size and the sharper its distribution in the membrane will be. The thickness of the layer L_g increases linearly with the square root of the dipping time. The process is quantitatively described by Leenaars and Burggraaf (1985). The rate of membrane deposition increases with the slip concentration or with decreasing pore size of the support as shown below. This has been experimentally confirmed for alumina and titania (Leenaars and Burggraaf 1985, Uhlhorn et al. 1989).

The capillary pressure drop, ΔP_c, caused by pores with an effective radius r for each capillary is given by

$$\Delta P_c = (2\tau/r)\cos\theta \qquad (2.1)$$

where τ = surface tension, θ = contact angle between liquid and support, and

$$L_g = \left(\frac{2K_g \Delta P_g t}{\eta\alpha}\right)^{1/2} \qquad (2.2)$$

where L_g is the permeability constant of the gel layer, η the viscosity of the slip "liquid", K_g, a constant related to the reciprocal of solid concentration and ΔP_g the pressure drop across the gel layer. ΔP_g can be eliminated from Equation 2.2 and expressed in terms of the above mentioned parameters.

Tiller and Chum-Dar Tsai (1986) discuss the theory of slip-casting in general and they show that there is an optimum pore diameter for producing a maximum pressure drop across the formed cake (this is the gel layer in the case of membrane formation) to give a maximum rate of cake formation. The results of both groups of investigators show the importance of support pore size and structure, and of the effects of the gel layer structure which is incorporated in the value of K_g. The value of K_g can be expressed in terms of structural constants of the membrane if the structure is known or if a model is assumed (Leenaars 1984, 1985). After the gel layer is formed it is dried. This is a very critical process step because large capillary forces are set up during the removal of the liquid. A xerogel layer is formed and large stresses due to shrinkage along the depth of the membrane occur which have to be released in some way. If a critical stress is exceeded cracks are formed in *supported* membranes. This occurs at a critical thickness (1–10 μm) of the membrane which strongly depends on the forming conditions and on the morphology of the material (plates or spheres) and the support. The effect of the support is demonstrated by the fact that *nonsupported* membranes can be produced in thicknesses up to 100 μm under room temperature and standard humidity conditions (about 60%). The compaction stresses can be used to order the compact structure during the relaxation process of these stresses and this contributes to a narrow pore size distribution (Leenaars 1984, 1985, van Praag et al. 1990). From the work of the group led by Burggraaf and Keizer, it emerges that the production of defect-free membranes is easier for plate-shaped particles (alumina) than for spheres (titania) (van Praag et al. 1990). From Equation 2.2, it can be seen that the viscosity of the slip plays an important role. It regulates the formation rate of the gel layer and helps to prevent the slip from penetrating the support pore system. In the colloidal suspension route the evolution of the viscosity during the solvent extraction is slow during the very first steps of the process and drastically increases just before gelling. With the polymeric gel route a more gradual increase of the viscosity is observed. In both cases the evolution of the viscosity can be modified by the addition of binders to the sol "slip". Different kind of binders are chosen depending on the nature of the solvent, the compatibility with the precursors and the viscosity of the system.

Finally, binders or plasticizers can play an important role in the prevention of cracks in the layer. As shown by van Praag et al. (1990), titania-supported membranes can be formed with plasticizers on normal supports, while without plasticizers these can be obtained only on special supports with very small roughness. Frequently used binders/plasticizers include polyvinylalcohols, cellulosic compounds and polyglycols in an aqueous medium and polyvinylbutyral in an alcoholic medium. It is important that the organic material can be completely pyrolyzed at relatively low temperature without leaving carbon or metal residues. After drying, the xerogel is first calcined to form an oxide structure. Further heat treatment strongly affects the final pore size of the membrane (Burggraaf, Keizer and van Hassel 1989, Leenaars et al. 1984, 1985, Larbot et al. 1988). Temperature and time also strongly influence the phase compositions (e.g. alumina, titania and zirconia membranes). At phase transitions (γ-θ-α alumina or anatase–rutile TiO_2) there is a strong increase in pore size (Larbot et al. 1987, Keizer and Burggraaf 1988).

It is obvious that the pore diameter can be regulated by heat treatment to values as small as 3–6 nm (minimum) and up to 50–200 nm depending on the material.

Although inorganic (ceramic) membranes offer many advantages they do suffer from a few limitations at the present state of technology development

Table 2.3. Advantages and Disadvantages of Inorganic (Ceramic) Membranes

Advantages

1. High temperature stability
2. Mechanical stability under large pressure gradients (noncompressible, no creep)
3. Chemical stability (especially in organic solvents)
4. No ageing, long lifetime
5. Rigorous cleaning operation allowable (steam sterilization, high backflush capability)
6. (Electro) catalytic and electrochemical activity easily realizable
7. High throughput volume and diminished fouling
8. Good control of pore dimension and pore size distribution

Disadvantages

1. Brittle character needs special configurations and supporting systems
2. Relatively high capital installation costs
3. Relatively high modification costs in case of defects
4. Sealing technology for high-temperature applications may be complicated

(see Table 2.3). Some typical synthesis methods and results for different ceramic materials will be discussed in the next section.

2.4. TYPICAL RESULTS FOR DIFFERENT MATERIALS

In the discussions to follow two types of membranes must be distinguished: (1) nonsupported and (2) supported ones. Nonsupported membranes are produced by pouring a slip onto a very smooth, dense substrate on which gelling takes place by slow evaporation of the dispersion liquid. In this way rather thick, crack-free films can be obtained. They are especially suitable for characterization purposes and structural investigations. These are evaluated to determine whether or not the structures obtained are similar or comparable with those of the supported ones, made from the same slip and the same material. The next section will also focus on the supported membranes.

2.4.1. Alumina Membranes

The mode of synthesis of alumina membranes through the colloidal suspension route is given in Figure 2.6. The first step involves the preparation of a slip consisting of boehmite particles. These are plate-shaped in the form of "pennies" with a diameter of 25–50 nm and a thickness of 3.5–5.5 nm (Leenaars et al. 1984, 1985). The synthesis chemistry of the colloidal boehmite (γ-AlOOH) solution is described in detail by Leenaars and Yoldas (1975) and to some extent by Anderson, Gieselman and Xu (1988) and by Larbot et al. (1987).

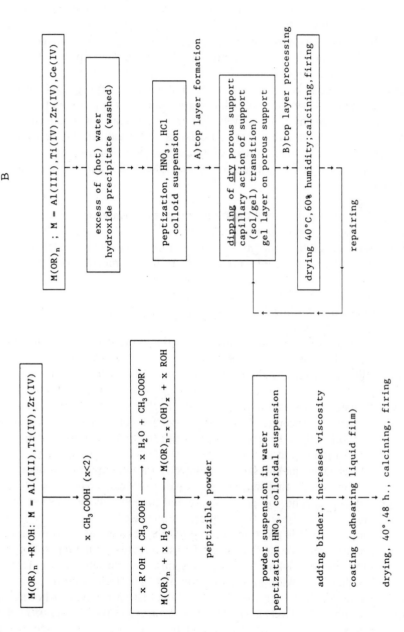

Figure 2.6. Synthesis route of alumina (a) the colloidal suspension route and (b) the slip-casting process (Larbot et al. 1987, Uhlhorn et al. 1989).

The process starts with the controlled hydrolysis of aluminum tri-sec butoxide (ATSB) or its dilute solution in 2-butanol in excess water at 80–90°C. At lower temperatures other aluminum compounds are formed which are not easily transformed into colloidal solutions and gels. After removing the alcohol from the mixture the precipitate is peptized with an appropriate amount of acid at 90–100°C at a pH value lower than 1.1 (Larbot et al. 1987), and in our studies at a pH of about 3.5. Supported gel layers are now formed by letting the colloidal solution come in contact with the porous support (slip-casting process). Whether or not a gel layer will be formed and what the casting rate (increase of thickness with time) will be, depend in a complex way on a large number of parameters. Some of the most important ones are investigated by Leenaars et al. (1984, 1985) and are summarized below

1. At a given pore size of the support a certain minimum concentration is necessary to obtain a gel layer, otherwise "pore clogging" will occur. With an increase in the sol concentration the casting rate increases. Typical concentrations used are from 0.7–1.2 mol AlOOH/L (see (5) below).

2. In the peptizing acid series (boehmite sols) the gelling concentrations increase according to the order $HCl > HNO_3 > HClO_4$ (gelling volume decreases). This implies an increasingly dense gel and a decreasing casting rate in the same order as in this series. This has been experimentally verified.

3. With an increasing quantity of acid used per mole AlOOH the mean pore size of the membranes after calcination is slightly decreased. Probably the stacking density of the particles in the gel increases with increasing concentration. Typical amounts of acid used range from 0.05–0.1 mol/L. Anderson's work seems to confirm this type of dependence (Anderson et al. 1988).

4. The gel layer thickness increases linearly with the square root of dipping time indicating that indeed a slip-casting process is operative. The rate constant depends on gel structure and pore size of the support. If the modal pore size of the support is increased from 0.12 μm (type 1 support) to 0.34 μm (type 2 support) the casting rate is decreased in accordance with theory. Typical casting rates for type 1 and type 2 supports are 4.4 μm/s$^{1/2}$ and 2.8 μm/s$^{1/2}$, respectively for HNO_3-stabilized sols with a concentration of 1.22 mol boehmite/L.

5. Ageing of the sol profoundly affects the casting behavior. After ageing (e.g. one week) gel layers could be formed on type 2 supports while before ageing this was not found to be the case. This points to an increase of the agglomerate size of the boehmite particles in accordance

with light scattering experiments of Ramsay, Daish and Wright (1978). Experiments with different acids suggest an increasing agglomerate size in the peptizing acid sequence $HCl > HNO_3 > HClO_4$.

In all cases the number of "pinholes" and of other types of casting defects is critically dependent on the quality of the support. Even in cases where the same nominal support material is used (but from different batches) varying results are obtained. This sensitivity of support quality could be diminished by adding an organic additive. In our experiments we used polyvinylalcohol, PVA, with a molecular weight of 72,000 and of the type giving a very low residue of ash or tar on pyrolysis. A typical standard "solution" contains 0.6 mol $AlOOH/L$ (peptized with 0.07 mole HNO_3 per mole $AlOOH$) with about 25–30 wt.% PVA based on dry Al_2O_3 (or 20 wt.% based on $AlOOH$).

As discussed by Larbot et al. (1987) and Cot, Guizard and Larbot (1988) the addition of organic "binders" has a dual function. First it regulates (increases) the viscosity of the solution and prevents the solution from being sucked into the pores of the support before gelation starts. It promotes a more gradual increase in the viscosity during the gelation process and thus makes the (initial) steps of the slip-casting process less critical. This, however, changes the rate constant of this process. A second function is the diminished tendency for crack formation during the drying (and subsequent calcination) steps. The wet gel layers must be dried under carefully controlled conditions which are typically 3 h at 40°C in an atmosphere of 60% relative humidity (Burggraaf, Keizer and van Hassel 1989) or 48 h at 40°C (Larbot et al. 1987). This step is very important, because inadequate drying results in cracked layers. This can be understood from the very large capillary and shrinkage stresses formed within the layer. (Scherer 1986, 1987). Imbalances in these stresses must undergo relaxation to some extent to prevent crack formation. At the same time these stresses are responsible for the ordering of the initially randomly packed, particles or agglomerates. In the case of plate-shaped particles, as with γ-$AlOOH$, the final result is represented in Figure 2.7. It can be easily seen that in this case the pores are slit-shaped. The critical diameter of the pore is related to the thickness of the boehmite plates (of about 3 nm) with a narrow pore size distribution. Experimental confirmation of ordering during drying has been given by Leenaars and Burggraaf (1985) and is based on the following main observations on nonsupported membranes:

1. A strong texture can be deduced from the XRD spectrum. The intensity of the [020] reflection as compared with those of the [120] and [031] reflections in boehmite is 10 times larger than can be expected for a random distribution of crystal directions and is in accordance with the model shown in Figure 2.7.

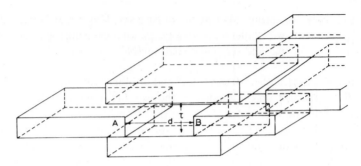

Figure 2.7. Idealized model of the boehmite membrane structure. d is the distance between the 2 boehmite crystals A and B, τ is the thickness of the boehmite plates (Leenaars and Burggraaf 1985).

2. Drying without the occurrence of large capillary stresses was obtained with supercritical drying in an autoclave. In this case a mean pore size was obtained which was twice that obtained under normal drying conditions and with a broad pore size distribution in accordance with the expectation for a noncompressed, random packing of particles.

There exist a maximum allowable thickness of the supported gel layers above which it is not possible to obtain crack-free membranes after calcination. For γ-alumina membranes this thickness depends on a number of (partly unknown) parameters and has a value between 5 and 10 μm. One of the important parameters is certainly the roughness and porosity of the support system, because unsupported membranes (cast on teflon) are obtained crack-free up to 100 μm. The xerogel obtained after drying was calcined over a wide range of temperatures. At 390°C the transition of boehmite to γ-Al$_2$O$_3$ takes place in accordance with the overall reaction

$$2\gamma\text{-AlOOH} = \gamma\text{-Al}_2O_3 + H_2O\uparrow \qquad (2.3)$$

This transition produces an isomorphous phase and the resulting γ-alumina has the same morphology and texture as its boehmite precursor. With increasing temperature and time the mean pore diameter increases gradually and other phases appear (δ-, θ-alumina). Due to the broad XRD lines, the distinction between γ- and δ-alumina cannot be made; θ-alumina occurs at about 900°C while the conversion to the chemically very stable α-alumina phase takes place at $T > 1000$°C. Some typical results for alumina membranes synthesized without binders are given in Table 2.4. When PVA was used as a binder, thermogravimetric analysis showed that, provided the appropriate binder type was used, the binder could be effectively removed at $T \geqslant 400$°C. The ash residue is of the order of 0.01 wt.%. Mean pore size and

Table 2.4. Microstructural Characteristics of Alumina Membranes as a Function of Calcination Time and Temperature (Leenaars et al. 1984, Burggraaf et al. 1989, Uhlhorn et al. 1989b)

Temperature (°C)	Time (h)	Phase	BET Surface Area (m²/g)	Pore Size (nm)	Porosity (%)
200	34	γ-AlOOH	315	2.5*	41
400	34	γ-Al$_2$O$_3$	301	2.7	53
	170	γ-Al$_2$O$_3$	276	2.9	53
	850	γ-Al$_2$O$_3$	249	3.1	53
500	34	γ-Al$_2$O$_3$	240	3.2	54
700	5	γ/δ-Al$_2$O$_3$	207	3.2	51
	120	γ/δ-Al$_2$O$_3$	159	3.8	51
	930	γ/δ-Al$_2$O$_3$	149	4.3	51
800	34	γ/δ-Al$_2$O$_3$	154	4.8	55
900	34	θ-Al$_2$O$_3$	99	5.4	48
1000	34	α-Al$_2$O$_3$	15	78[†]	41
550	34	γ-Al$_2$O$_3$	147	6.1[‡]	59

* All pore sizes are according to the slit-shaped model
[†] Cylinder-shaped pore model (diameter)
[‡] Prepared from a sol treated in an autoclave at 200°C

pore size distribution did not change on addition of the binder, the porosity however increased somewhat.

It can be observed from Table 2.4 that the transition to α-alumina is accompanied by a large increase in pore diameter. Results of Larbot et al. (1987) show the same trend but larger pore diameters with more pronounced increase with temperature was observed (e.g. 10 nm at 900°C). The rate of increase of pore diameter with temperature at the γ-θ-α transitions was, however, smaller (e.g. 25 nm at 1100°C for α-alumina membrane).

Thus, thermally stable alumina membranes can be produced with a pore diameter as low as 3 nm. For long-term thermal stability the temperature of heat treatment should be 50–100°C higher than the applied temperature. Typical pore size distributions, e.g. for γ-alumina membrane, are given in Figure 2.8. If the desorption branch of adsorption/desorption isotherms is used this yields the smallest passage in a packed array. This is, however, exactly what is known as "effective" pore diameter in membranes. If for alumina membranes a slit-shaped pore model is used for the calculations, a good match is obtained between experimentally found values for pore size (distribution), porosity and BET values and those calculated from particle size (XRD, electron microscopy) and model representations (Figure 2.7). This implies that the agglomerates which are present in the colloidal suspension

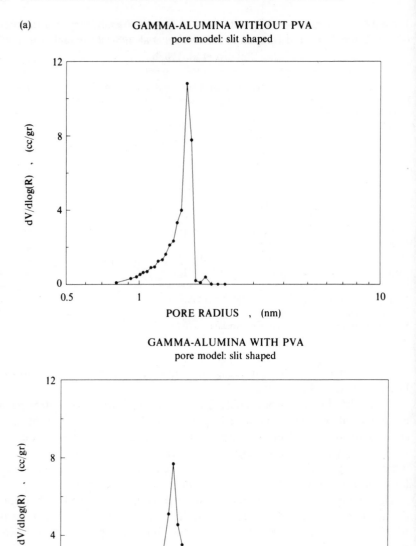

Figure 2.8. Typical examples of a pore size distribution for (a) γ-alumina membranes; desorption branch (b) anatase titania membranes; desorption branch.

(b)

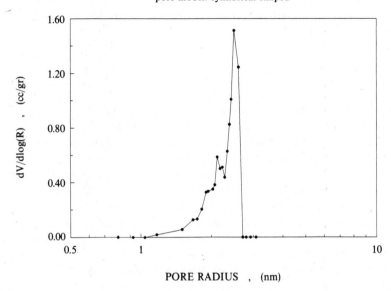

TITANIA WITHOUT BINDERS
pore model: cylindrical shaped

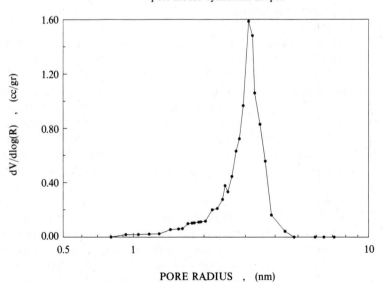

TITANIA WITH PVA AND HPC (Mw = 10E5)
pore model: cylindrical shaped

Figure 2.8 (continued)

are "collapsed" during the type of drying process in which large capillary pressures are used. Cot, Guizard and Larbot (1988) used agglomerate size as a parameter to obtain larger pores for microfiltration membranes (see Sections 2.4.2 and 2.4.3 for zirconia and titania respectively).

2.4.2. Zirconia Membranes

Zirconia membranes on carbon supports were originally developed by Union Carbide. Ultrafiltration membranes are commercially available now under trade names like Ucarsep and Carbosep. Their outstanding quality is their high chemical resistance which allows steam sterilization and cleaning procedures in the pH range 0–14 at temperatures up to 80°C. These systems consist of a sintered carbon tube with an ultrafiltration layer of a metallic oxide, usually zirconia. Typical tube dimensions are 10 mm (outer diameter) with a wall thickness of 2 mm (Gerster and Veyre 1985).

The fabrication of the ultrafiltration top layer is described in the patent of Cacciola and Leung (1981). The coating is applied to the macroporous carbon support in the presence of a volatile liquid capable of drawing the coating into the support and of desolvating the coating. The liquid is thereafter volatilized with the resulting dry ultrafiltration membrane being crack-free and having good mechanical and chemical stability. The support may be pretreated with a volatile liquid such as acetone, prior to the application of the coating to the support. Alternatively, the coating material can be dispersed in a volatile liquid, such as methanol, to form a coating suspension that is applied to the support.

The coating material, preferably zirconia, will have a primary particle size in the range 20–150 nm and an aggregate size in the range 100–1000 nm. The coating thickness is about 20 μm. The carbon tube support has a pore size of about 300 nm and a porosity of about 30%. Stability of the coating is enhanced by firing at temperatures of 400–600°C. The separation properties depend on the applied primary particle size and pores of 5 nm and larger can be obtained (Veyre 1985). This type of ultrafiltration membrane is commercially available and is produced by Tech Sep (formerly SFEC) under the name of "Carbosep". The applications are in the field of treatment of water streams (oily wastes) (Bansal 1975, 1976) and food and dairy industry (e.g. cheese production, Veyre 1985a, b).

Guizard et al. (1986), Cot, Guizard and Larbot (1988) and Larbot et al. (1989) used a sol–gel method to prepare zirconia membrane top layers on an alumina support. The water necessary for the hydrolysis of the Zr-alkoxide was obtained from an esterification reaction. The complete hydrolysis was done at room temperature and resulted in a hydrated oxide. The precipitate was peptized with nitric or hydrochloric acid at pH < 1.1 and the final sol contained about 20% of metal oxide. Important parameters for the sol–gel transition are the pH and the concentration of the solution. In order to give a compact arrangement after gelation the particles should be in a maximum repulsion state (Cot, Guizard and Larbot 1988). With a pH near the isoelectric point, larger particles are formed and poorly densified porous films result. A similar effect is obtained by increasing the electrolyte concentration in the solution. Cot, Guizard and Larbot (1988) report particle sizes of 10 nm for a low electrolyte concentration and 60 nm for an electrolyte concentration which is 100 times higher. This particle size increases further if the ZrO_2 concentration in the solution is increased above 1%. The particles are agglomerates whose sizes can be manipulated. After addition of a binder the zirconia membrane was formed on an alumina support using the slip-casting process. After drying (20–150°C) and thermal treatment (400–900°C) the resulting membranes showed a broad pore size distribution. The zirconia membrane layer exists in the tetragonal phase up to 700°C, above which a phase transition to the monoclinic form occurs. The pore diameter increases gradually with increasing temperature from 6 nm (700°C) to 70 nm (1200°C). Results given by Cot, Guizard and Larbot (1988) show a curve with a mean pore radius of 32 nm with considerable "tails" with smaller and larger values. It appears that the membranes are prepared from rather large agglomerates. It is questionable if these membranes are crack- and defect-free (see Section 2.10).

2.4.3. Titania Membranes

Supported titania membranes are described by several authors (Burggraaf et al. 1989, Zaspalis et al. 1989, Cot, Guizard and Larbot 1988), while the preparation of nonsupported and supported titania membranes is reported by Anderson et al. (1988) and Gieselman et al. (1988). Titania membranes show excellent chemical resistance and interesting photochemical and catalytic properties. The results of titania membranes reported by Cot, Guizard and Larbot (1988) were obtained in the same way as described for zirconia. Particle diameters ranging from 20–170 nm are reported, depending on preparation conditions (Cot, Guizard and Larbot 1988) and the same type of broad pore size distribution as reported for zirconia. Up to 500°C the material is present in the anatase phase, above which it is transformed to the rutile phase. The pore diameter increased gradually with increasing temperature from 6 nm (500°C) to 180 nm (1100°C). Again it is doubtful if these membranes are defect-(crack) free (see below).

The results obtained by Burggraaf, Keizer and van Hassel (1989a, b) differ markedly from those reported above by Cot, Guizard and Larbot (1988). The preparation of defect-free titania membranes was found to be much more difficult than that of alumina membranes. It was necessary to use binders/plasticizers and better results were obtained with very smooth support surfaces. Sols were obtained by hydrolysis of Ti-tetraisopropoxide dissolved in isopropanol (0.45 M) at room temperature in an excess of a water/isopropanol mixture (4.5 M water in isopropanol) to which a small amount of sulfuric acid is added.

Peptization of the precipitate was obtained with nitric acid at a pH of 1.5 at 70°C under reflux conditions. At pH values higher than 3, peptization was not obtained. The peptized sols consisted of stable colloidal dispersion. Using light scattering technique agglomerate sizes of about 26 nm were observed which after ageing for two months increased to about 46 nm. On subjecting the solution to ultrasound the agglomerate size decreased to about 9 nm. Transmission electron microscopy showed primary particle sizes of 4–5 nm (van Praag et al. 1990). Titania membranes were produced on alumina supports from colloidal suspensions containing 0.1–0.2 mol/L TiO_2 after addition of PVA (polyvinylalcohol) or a combination of PVA with hydroxypropylcellulose (HPC). A layer thickness of about 1 μm was deposited followed by controlled drying and calcination at 450°C. This procedure is repeated until the required thickness is obtained and/or a defect-free membrane is obtained as indicated by gas permeation measurements. Nonsupported membranes could be produced in this way with a thickness up to 50 μm. The addition of SO_4^{2-} ions was necessary to stabilize the anatase phase up to a temperature of 600°C. Without this addition the anatase to rutile phase transition takes place at 350–450°C. This is accompanied by a decrease of the unit cell volume by 8% and probably gives rise to cracking.

Characteristic microstructural properties of TiO_2 membranes produced in this way are given in Table 2.5. Mean pore diameters of 4–5 nm were obtained after heat treatment at $T < 500$°C. The pore size distribution was narrow in this case and the particle size in the membrane layer was about 5 nm. Anderson et al. (1988) discuss sol/gel chemistry and the formation of nonsupported titania membranes using the colloidal suspension synthesis of the type mentioned above. The particle size in the colloidal dispersion increased with the H/Ti ratio from 80 nm ($H^+/Ti^{4+} = 0.4$, minimum gelling volume) to 140 nm ($H^+/Ti^{4+} = 1.0$). The membranes, thus prepared, had microstructural characteristics similar to those reported in Table 2.5 and are composed mainly of 20 nm anatase particles. Considerable problems were encountered in membrane synthesis with the polymeric gel route. Anderson et al. (1988) report that clear polymeric sols without precipitates could be produced using initial water concentrations up to 16 mole per mole Ti. Transparent gels could be obtained only when the molar ratio of H_2O to Ti^{4+} is $\leqslant 4$. Gels with up to 12 wt.% TiO_2 could be produced provided a low pH is used ($H^+/Ti^{4+} \leqslant 0.025$).

Table 2.5. Microstructural Characteristics of Titania and Ceria Membranes as a Function of Calcination Time and Temperature (Uhlhorn et al. 1988, Burggraaf, Keizer and van Hassel 1989a, b)

Membrane Material	Time (h)	Temperature (°C)	Pore Diameter (nm)	Porosity (%)	BET Surface Area (m²/g)	Crystallite Size (nm)
TiO_2[†]	3	300	3.8*	30	119	Sol: 5[‖]
						Sol: 10–40**
TiO_2[‡]	3	400	4.6*	30	87	
TiO_2[§]	3	450	3.8*	22	80	
TiO_2[§]	3	600	20*	21	10	
CeO_2	3	300	≈ 2	15	41	Sol: 10 (TEM)
CeO_2	3	400	≈ 2	5	11	
CeO_2	3	600	nd[#]	1	1	

* Cylinder-shaped pore model
[†] Anatase phase
[‡] Rutile phase
[§] Titania stabilized with SO_4^{2-} ions
[‖] Primary particle (TEM)
[#] nd is not detectable
** Agglomerated sol particles (10 nm: use of ultrasonic waves, laser scattering)

2.4.4. Silica Membranes

The synthesis of silica membranes has only recently been described. Silica forms sols and gels very easily both by the colloidal suspension and by the polymeric gel route. Its chemical resistance and its thermal stability in the presence of water vapor or metal impurities are not very good however. Larbot et al. (1989) have described the synthesis of silica membranes starting with a commercially available silica sol (Cecasol Sobret) in an aqueous solution at pH 8.

A mixture of colloidal silica (20 wt.% SiO_2), water and 5 wt.% organic additives/cellulosic type as binder, with glycol as plasticizer was used to deposit a SiO_2 film on alumina porous substrates (0.2 μm pores). The deposited films were dried. This step determines the thickness, particle size, pore size and porosity of the resulting membranes. Organic additives were removed by heat treatment in air at 200°C. Finally heat treatment at 400–800°C was applied. Surprisingly Larbot reports that the initial thickness of the membrane greatly influences the final particle size. For a membrane of 5 μm thickness the pore diameter could be varied between 6.0 and 10.0 nm depending mainly on the drying temperature. This was explained by the residual concentration of water (vapor) present in the subsequent calcination step. Larger water concentration results in stronger sintering (loss of surface area). The calcination temperature did not affect the pore size in an important way up to a temperature of 700–800°C but prolonged (10 h) heat treatment at 900°C produced a sintered coating with closed pores. The concentration of hydroxyl groups at the surface decreases with increasing calcination temperature. Kaiser and Schmidt (1984) synthesized silica membranes with a thickness of 3–10 μm by reacting a porous glass fiber support filled with water and a catalyst (2–6 mol HCl/L) with alkoxy silane [$Si(OCH_3)_4$] vapor at 45°C. They report a resulting mean pore diameter below 2.5 nm without further discussion of the results. Finally Klein and Gallagher (1988) and Gallagher and Klein (1986) report the synthesis of thick (100 μm) nonsupported sheets of silica with large pores (30 μm). Interesting results with silica have been obtained in membrane modification processes and will be discussed in Section 2.9.

2.4.5. Binary Composite Membranes

RuO_2–TiO_2 Guizard et al. (1986a, 1989) report the synthesis of electrically conductive RuO_2/TiO_2 membranes on alumina supports by both the colloidal suspension and the polymeric gel route. Precursors were $Ti(OPr)_4$ and $RuCl_4 \cdot 3H_2O$ dissolved in propanol (2–10%). Only a few details are given. In the polymeric gel route the hydrolysis reaction was controlled by esterification of the appropriate alcohol (Guizard et al. 1986a). After peptization of 5–10% powder suspension in water with nitric acid at 80–90°C, binders were added and a gel layer was formed on the support. Organic products were removed at 380°C. In the region 400–700°C crystallinity develops more rapidly in the colloidal suspension gel as compared to that observed in the polymeric gel. Initially (400°C) a mixture of tetragonal RuO_2 and anatase (TiO_2) is present, the latter being gradually converted to the rutile phase at higher temperature. Mean pore diameters of 10–20 nm for colloidal gels and of 5 nm for polymeric gels were reported. The pore size distribution showed a narrow peak with a large "tail" towards larger pore sizes.

Al_2O_3–TiO_2, Al_2O_3–CeO_2 and Al_2O_3–ZrO_2 binary systems   Uhlhorn et al. (1988) report the synthesis of nonsupported membranes consisting of mixtures of plate-shaped γ-alumina particles with spherically-shaped TiO_2, CeO_2 or ZrO_2. In the case of alumina–titania, sols with a concentration of 1 M boehmite and 0.1 M titania were produced by the colloidal suspension route as described earlier. At pH values below 5 both sols can be mixed without destabilization. The concentration of alumina in the final slip-cast dispersion must be larger than 0.7 M for layer formation to occur. PVA was used as a binder (3–4 wt.% on metal basis). Binary membranes were formed

Table 2.6. Microstructural Characteristics of Some (Nonsupported) Binary Composite Membrane Systems (Uhlhorn et al. 1988, Burggraaf, Keizer and van Hassel 1989a, b)

Membrane Material	Time (h)	Temperature (°C)	Pore Diameter (nm)	Porosity (%)	BET Surface Area (m^2/g)	Crystallite Size (nm)
Al₂O₃/CeO₂ (35 wt.%)	3	450	2.4*	39	164	Sol: 10 and 50† Calc.: 5 (CeO₂)
Al₂O₃/CeO₂ (35 wt.%)	3	600	2.6	46	133	
Al₂O₃/TiO₂ (30 wt.%)	3	450	2.5	48	260	
Al₂O₃/TiO₂ (55 wt.%)	3	450	2.5	38	220	

All pore sizes are according to a slit-shaped pore model
Bimodal distribution of CeO₂ particles

by slip-casting the dispersion on α-alumina supports or on supported γ-alumina membranes. Similarly alumina/ceria membranes were made.

Characteristic properties of nonsupported layers of some binary combinations heat treated at different temperatures are given in Table 2.6. Interesting features are that (1) the synthesis of crack-free layers in binary systems with a major phase of alumina is easier than with single phase titania, (2) the crystal growth of the minor phase is strongly retarded (compare CeO₂ particles of 5 nm after calcination at 450°C), whereas sintering of single-phase CeO₂ membranes always resulted in dense coatings and (3) the pore characteristics of binary mixed oxide membranes are comparable with those of the γ-alumina membranes.

The synthesis of 5 μm thick TiO_2–SiO_2 layers on a porous support can be performed using the procedure given below. First a mixed $Ti[(OMe)_3]_4$ alkoxide is synthesized by reacting partially hydrolyzed $Si(OMe)_4$ with Ti-isopropoxide. This inorganic polymer is hydrolyzed at pH 11.0 and treated with 2-methyl-2-4-pentanediol and a binder. This solution is then slip-cast onto a porous support, dried and calcined at 700°C. The membrane can be useful in reverse osmosis applications.

2.5. PHASE SEPARATION/LEACHING METHODS AND GLASS MEMBRANES

Porous glass membranes are produced by leaching a two-phase glass body in the system Na_2O–B_2O_3–SiO_2 (Schnabel and Vaulont 1978, Schnabel 1976).

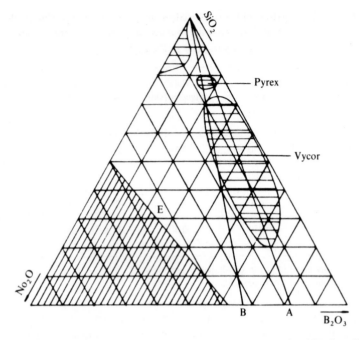

Figure 2.9. Phase diagram of the system $Na_2O-B_2O_3-SiO_2$ (Schnabel and Vaulont 1978).

The classical composition is 70 wt.% SiO_2, 23 wt.% B_2O_3 and 7 wt.% Na_2O reported in the investigations of Hood and Nordberg in the late thirties and early forties (e.g. Hood and Nordberg 1938). The two-phase region (miscibility gap) in the phase diagram (Figure 2.9) shows, however, a larger composition range. This miscibility gap is known as Vycor.

Under certain time and temperature conditions, the homogeneous glass separates into two phases. One of the phases consists substantially of silicon dioxide which is insoluble in mineral acid. The other phase represents a soluble coherent boric acid phase rich in alkali borate. If the boric acid phase is dissolved out of this heterogeneous glass structure with a mineral acid, a porous skeleton of substantially insoluble silicon dioxide is left. The phase separation region occurs between 500°C and 800°C.

The homogeneous glass, in which a few percent of alumina is added for better processing, is prepared from a melt at 1300–1500°C (Schnabel 1976) or 1000–1200°C/1450°C (McMillan 1980). It is important for the properties that the melt is as homogeneous as possible. Schnabel (1976, 1978) produced glass capillaries or hollow fibers directly from the glass melt at a viscosity of 10^3 P. The phase separation was carried out by heat treatment between 500–800°C.

A visible sign for the phase separation is that glass exhibits increasing opalescence with increasing temperature. Because the pore sizes are regulated with the annealing temperature (at a constant time and composition) glass becomes completely opaque near the temperature limit of the miscibility gap (800°C).

Also the annealing time influences the phase separation. As a consequence there are many possibilities of producing porous glass bodies by changing glass composition, annealing time and temperature. Furthermore, care must be taken that the glass melt is directly cooled to the glass temperature, otherwise prephase separation occurs due to the temperature gradient resulting in an inhomogeneous pore distribution. If proper precautions are taken, a very narrow pore size distribution can be obtained after leaching. Leaching is carried out, for example, with 3–5 mol (HCl–KCl) solution in water at a temperature of 90°C (Schnabel 1976). The pore size distribution of a glass

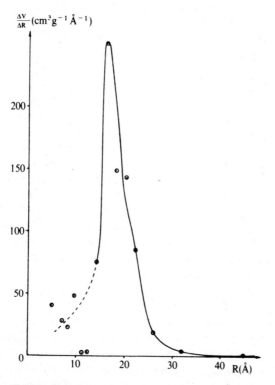

Figure 2.10. Pore radius distribution curve for a porous Vycor glass heat-treated at 500°C for 5 h (McMillan 1980).

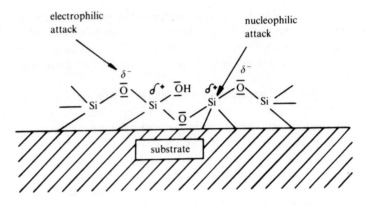

Figure 2.11. A schematic of glass surface with silanol groups (Schnabel and Vaulont 1978).

membrane annealed at 500°C measured with adsorption/desorption technique is shown in Figure 2.10. The measured pore diameter was 3–4 nm with a porosity of 20–30% (McMillan 1980). Pore diameters can change between the above mentioned value and 200 nm depending on the annealing treatment. The porosity increases with increasing pore size.

The main disadvantage of porous glass membranes is the instability of the surface. Thermal instability is caused by a further demixing at temperatures higher than the annealing temperature. Also the internal glass surface is rather active mainly due to the presence of silanol groups (Schnabel and Vaulont 1978). In Figure 2.11 such a glass surface is shown. It is quite clear that a material such as porous glass cannot be considered as an "inert" material. By chemical modification and by introduction of any desired functional group the physical, chemical and thermal properties of the membrane surface can be changed. A summary of possible surface reactions and surface modifications is shown in Figure 2.12. Because of its hydrolytic instability the formation of ≡Si–C≡ -bond is the most important reaction (reactions 1, 2, 12, 13, and 14). More sensitive but essential for the production of intermediary products in the silicon and organosilane chemistry is the ≡Si–O–C≡ -bond, represented in the reactions 3, 4, 5, 6, 7, 8, and 10. The reactions 3 and 10 are often called "surface esterification". The replacement of surface silanol groups can be attained by halogen atoms either by fluorination (reaction 9) or by chlorination. Unlike the quite stable ≡Si–F -group, the surface ≡Si–Cl -bond is unstable and further reactions may occur (reactions 10, 11, 12, 13, 14). These few examples may serve as evidence to show that the surface silanol group can react chemically in the same manner as other functional groups. Due to the reactivity of this silanol group a wide range of organic reactions can be performed on a glass surface. This offers the

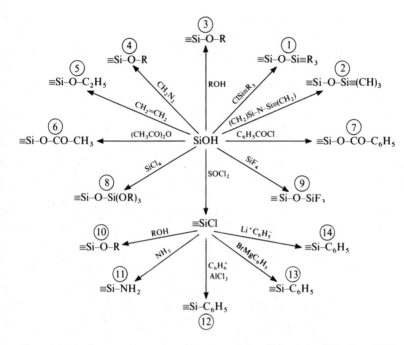

Figure 2.12. Surface reactions and modifications of glass (Schnabel and Vaulont 1978).

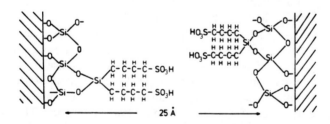

Figure 2.13. Pore and wall structure of a modified membrane suitable as desalting membrane for RO applications (Schnabel and Vaulont 1978).

possibility of "tailor-made" surfaces for specific applications. A schematic drawing of a modified membrane is presented in Figure 2.13 for desalination applications. This type of membrane can reject NaCl from a 0.5% NaCl solution in water, at 100 bar with a retention of 99% (Schnabel and Vaulont 1978). Unmodified systems show retention of 0–40% (Kraus et al. 1966), 0–70% (Elmer 1978), 80% (Schnabel and Vaulont 1978) or 90% (McMillan and Maddison 1981). Elmer concluded that desalination was also affected by siliceous deposits in the pores of the glass. These membranes are also suitable in blood purification (Baeyer et al. 1980) or protein separation (Schnabel, Langer and Breitenbach 1988). In these cases the surface modification is of primary importance, because the hydrophilic/hydrophobic activity can be changed. This affects interaction and separation.

Porous glass membranes (unmodified) were used for the separation of H_2/CO gas mixture utilizing the Knudsen diffusion mechanism (see Chapter 6.7 and Haraya et al. 1986) and also for gas separation using the membrane reactor concept (Itoh et al. 1988). Hammel et al. (1987) describe the manufacture of porous siliceous fibers and their gas separation properties. These hollow fibers with an outer diameter of 30–40 μm were used to separate a He/CH_4 mixture with a selectivity of 1000 and an O_2/N_2 mixture with a selectivity of 1.75. It is not clear how such large selectivities are related to the structure of these membranes. It is possible that a lower sintering temperature (400°C) gives a microporous siliceous, spongy system (pore diameter < 2 nm) with other gas transport properties. Such a system is not stable at temperatures higher than 400°C (see also Elmer 1978).

Thermally stable systems can be obtained by coating the internal pores with other inorganic components (McMillan and Maddison 1981). The porous glass is first treated with a metal chloride ($TiCl_4$) and the resulting surface layer is hydrolyzed to produce the metal oxide coating. The reaction of $TiCl_4$ with the porous glass is carried out by introducing 2 ml of the volatile liquid chloride into a N_2 stream. This gas mixture is brought into contact with the membrane (held at 80°C) and the chloride reacts with the chemically bonded hydroxyl group remaining on the glass surface. Other chlorides ($SnCl_4$) are possible but the treatment requires higher temperatures (400°C). The pore size is slightly decreased (from 2.5 to 2.2 nm) and a further reduction in pore size (1.8 nm) with an accompanying increase in strength can be obtained by partially sintering the treated membrane at 800°C. Eguchi et al. (1987) manufactured porous glass tubes (diameter 10 mm, wall thickness 1 mm) of the type SiO_2–ZrO_2 using a similar leaching method. The sintering temperature was between 625 and 740°C and the resulting pore size between 20 and 2000 nm indicating that smaller pores could not be achieved.

Finally, porous glass membranes were prepared on a porous alumina ceramic tube (Ohya 1986). Porous thin-glass membranes with a thickness of 5–23 μm have been prepared on a surface of a porous ceramic tube by coating with a metal alkoxide solution (e.g. $Si(OC_2H_5)_4 + H_3BO_3$ or B_2O_3 in ethanol/water/HCl), drying, heating and acid leaching sequentially. The pore size may range from 4 to 64 nm.

Porous glass membranes prepared by leaching methods can be obtained with pore size between 2 and 200 nm. The internal surface is active and can be treated/modified for special applications or to obtain higher chemical and thermal stability. Porous glass membranes show unique properties and may serve as suitable candidates in several potential applications.

2.6. ANODIC OXIDATION

The process of anodic oxidation of aluminum and the resulting structures which have been realized was described 30 years ago (Hoar and Mott 1959). Until now no other metal could be anodized successfully to obtain the same type of porous structures. The process for producing an anodic aluminum oxide membrane is described and patented by Smith (1973, 1974) and anodic aluminum oxide membranes are produced since 1986 on laboratory scale by Anotec Separations (1986). In Figure 2.14a, b the structure of this type of membranes is shown as reported by Anotec Separations (1986). Two types of membranes in laboratory-scale modules are available; (1) a homogeneous membrane with pores of 200 nm and a porosity of 65% or higher; (2) an asymmetric membrane with pores of 25 nm in the top layer. Both membranes have a thickness of 60 μm. The difference in structure is obtained by changing the process parameters in the preparation process. When aluminum is treated anodically in solutions of sulfuric, chromic and certain other acids an oxide film of the form shown in Figure 2.15 is formed. A compact film AB, of about the same thickness as that formed at the same bath voltage in nonacidic (e.g. borate) solutions, is formed initially, after which a much thicker porous film BC continues to grow while a portion of the oxidized metal appears as Al^{3+} ions in the acid solution. In the process, aluminum ions pass through the film and neutralize the ions formed at the interface between oxide and electrolyte. O'Sullivan and Wood (1970) made an extended study on the morphology and mechanism of formation of porous anodic films on aluminum. Structural studies were also carried out by Takahashi et al. (1973), Thompson et al. (1978), Furneaux, Thompson and Wood (1978) and Pavlovic and Ignatiev (1986). Siejka and Ortega (1977) used an ^{18}O study for field-assisted pore formation in anodic oxide films on aluminum. Some of the conclusions listed from the above studies are:

1. Pore initiation in anodic films produced at constant current density occurs by the merging of locally thickening oxide regions, which appear related to the substrate substructure, and the subsequent concentration of current into the residual thin regions.
2. The pores grow so that their diameter remains proportional to the applied voltage as the steady state is approached.
3. The barrier-layer thickness (AB in Figure 2.15), cell diameter and pore diameter are directly proportional to the formation voltage.

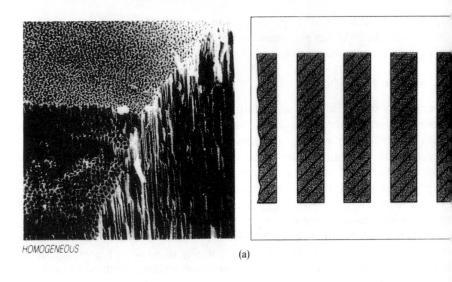

HOMOGENEOUS

(a)

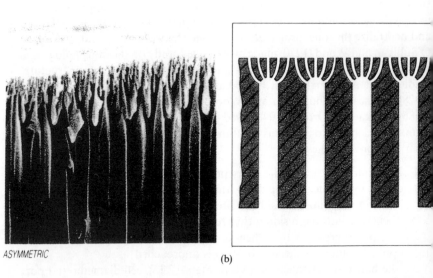

ASYMMETRIC

(b)

Figure 2.14. The structure of an anodized aluminum oxide membrane (Anopore) as shown in Anotec Separations (1986); (a) is a homogeneous membrane; (b) an asymmetric membrane.

Electrolyte

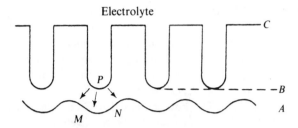

Figure 2.15. The formation of an oxide film on an aluminum film by anodic oxidation. Arrows show the path PM or PN of oxygen through the compact oxide film AB (barrier layer). BC is the already formed porous oxide film, M and N are located in the metal film (Hoar and Mott 1959).

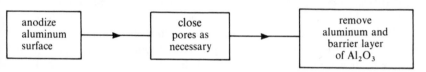

Figure 2.16. Production scheme of a anodic aluminum oxide membrane (Smith 1973, 1974).

4. Increase and decrease of the voltage during anodization leads to a redistribution of pore and cell populations, requiring pore merging or termination and pore initiation, respectively.
5. Use of relatively weak electrolytes produces porous films with thicker barrier layers, larger cells and larger pores next to the barrier layer than strong electrolytes under constant current density conditions.
6. Pore widening, film distortion and film collapse upon drying, produced by chemical dissolution during prolonged contact with the acid are more severe with strong electrolytes.

Smith (1973, 1974) patented the process for producing an anodic aluminum oxide membrane using the scheme shown in Figure 2.16. A sheet of aluminum is mounted in a cell in such a way that the sheet can be exposed to different solutions on both sides of the sheet. The preparation procedure is illustrated below: Aluminum foil (thickness 12 μm) was anodized by placing a solution of 15% H_2SO_4 (by weight) containing 1.5% $Na_2Cr_2O_7$ on one side of the foil. The anodization process was carried out at 15 V for a period of 37 min. At this time the foil was translucent. After rinsing, the sample was etched with a 15% solution of H_2SO_4 on the unanodized side and pure water on the anodized side (50 min). After rinsing again, the sample was placed in distilled water to hydrate for 20 min at 50–60°C and allowed to cool slowly.

The osmotic flow of the resultant membrane was determined by placing a 5 M NaCl solution on the unanodized side and pure water on the other side. A water permeability value of 3 mg/cm^2-h was obtained with a 3:1 water salt flow ratio. The flow and the separation could be improved by treatment with 15% H_2SO_4 and 1.5% $Na_2Cr_2O_7$ solution on the unanodized side and a moderately concentrated K_2CrO_4 solution (buffered in a basic medium) placed on the other side. The flow increased to 33 mg/cm^2-h which means that the barrier layer was not completely removed by the first etching. Further etching with 1 mol $Na_2Cr_2O_7$ and a buffered base produced a water flow of 110 mg/cm^2-h and a water salt flow ratio of 23:1. The process can be carried out with a higher voltage (50 V), a thicker membrane (50 μm) or with other combinations of acids. In this manner water/salt flow ratios up to 80:1 could be achieved.

Mitrovic and Knezic (1979) also prepared ultrafiltration and reverse osmosis membranes by this technique. Their membranes were etched in 5% oxalic acid. The membranes had pores of the order of 100 nm, but only about 1.5 nm in the residual barrier layer (layer AB in Figure 2.15). The pores in the barrier layer were unstable in water and the permeability decreased during the experiments. Complete dehydration of alumina or phase transformation to α-alumina was necessary to stabilize the pore structure. The resulting membranes were found unsuitable for reverse osmosis but suitable for ultrafiltration after removing the barrier layer. Beside reverse osmosis and ultrafiltration measurements, some gas permeability data have also been reported on this type of membranes (Itaya et al. 1984). The water flux through a 50 μm thick membrane is about 0.2 mL/cm^2-h with a N_2 flow about 6 cm^3/cm^2-min-bar. The gas transport through the membrane was due to Knudsen diffusion mechanism, which is inversely proportional to the square root of molecular mass.

Smith (1973, 1974) studied the problem of separating the remaining Al film from the anodic aluminum oxide film. Alcan (Furneaux, Rigby and Davidson 1985) patented a process to remove the metal film by reducing the applied voltage to a level preferably below 3 V (see also Furneaux, Rigby and Davidson 1989). A brief description of the procedure is as follows: A 99.98% aluminum panel, (5 × 5 cm) was chemically brightened and then anodized in 0.25 M oxalic acid at 25°C. For the anodizing procedure a current limit of 1.25 V and a voltage limit of 70 V were preset on the power supply. The resulting film was about 15 μm thick.

The voltage reduction was started immediately after completion of the anodizing process using the same electrolyte. The voltage was stepped down, in steps of 5% of the current voltage from the voltage at the end of the anodizing stage to about 0.1 V. Each time the voltage was reduced, the current fell to a very low value and then rose passing through a point of inflexion. The voltage reduction procedure took less than 1 h. The sample was then transferred to a 50% (by volume) H_3PO_4 solution and after 2–3 min small bubbles were observed at the metal/film interface. The sample was removed, well rinsed and dried in an oven. After drying, the film was detached from the metal. The oxide film had an asymmetric structure shown in Figure 2.2, the parallel pores had a diameter of about 250 nm; the pores in the barrier layer were about 10 nm and the thickness of the barrier layer was about 50 nm. The process can be repeated with other electrolytes using the principle of voltage reduction.

With anodic oxidation very controlled and narrow pore size distributions can be obtained. These membranes mounted in a small module may be suitable for ultrafiltration, gas separation with Knudsen diffusion and in biological applications. At present one of the main disadvantages is that the layer has to be supported by a separate layer to produce the complete membrane/support structure. Thus, presently applications are limited to laboratory-scale separations since large surface area modules of such membranes are unavailable.

2.7. PYROLYSIS

Although the pyrolysis of organic materials (organic hollow fibers) is used in the commercialization of a new family of inorganic membranes (Fleming 1988) there are only a few descriptions in the open literature. Koresh and Soffer (1980, 1986, 1987) have published a series of articles on this subject. There is also a paper by Bird and Trimm (1983) which is based on a previously described preparation procedure of Trimm and Cooper (1970, 1973).

Not only hydrocarbon systems, but also silicon rubbers (Lee 1986), can be pyrolyzed to obtain silicon-based membranes. Details of the pyrolysis are mainly reported for nonmembrane applications. A recent example is the paper of Boutique (1986) for the preparation of carbon fibers used in aeronautical or automobile constructions.

2.7.1. Molecular Sieve Carbons

In principle, molecular sieve carbons (MSC) can be achieved by the pyrolysis of thermosetting polymers such as polyvinylidene chloride, polyfurfuryl alcohol, cellulose, cellulose triacetate, polyacrylonitrile and phenol formaldehyde (Koresh 1980). An example is given by Trimm and Cooper (1970, 1973) for the preparation of MSC (mixed with metallic compounds) for catalyst systems. A mixture of furfuryl alcohol, platinum oxide and formaldehyde was heated to 40°C and additional formaldehyde was added to ensure the complete reduction of the oxide to the metallic state. The platinum produced was suspended in solution as a colloid. The solution was heated to 90°C and a solution of phosphoric acid was added to initiate polymerization of the alcohol. The polymer was cured at 110°C for 6 h, and then at 200°C for 6 h. The deposit was then crushed and carbonized at 600°C for 4 h. Such MSC can also be obtained without using platinum or another metal (i.e. by nonmetallic routes). Bird and Trimm (1983) used this method to prepare supported gas separation membranes. Of all the supports used (metals, carbon cloths and silica frits) only the silica frits gave some positive results in the separation of N_2/C_3H_6 and H_2/N_2.

Diffusivity ratios of 5 for N_2/C_3H_6 and 30 for H_2/N_2 at 350 K were obtained from steady-state measurements. These are in disagreement with the predictions (0.8 for N_2/C_3H_6 and 3.7 for N_2/H_2) based on the Knudsen diffusion mechanism. This is due to activated diffusion in the temperature range 290–450 K. In this work Trimm and Cooper (1970, 1973) concluded that these types of composite carbon membranes are difficult to prepare due to the large shrinkage of the carbon system during carbonization. However, in the opinion of the authors (Burggraaf and Keizer) this preparation method may be used with other types of supports (like carbon tubes or alumina systems with defined structures) and by using very thin layers (see also Fleming 1988).

A completely different approach was taken by Koresh and Soffer (1980, 1986, 1987). Their preparation procedure involves a polymeric system like polyacrylonitrile (PAN) in a certain configuration (e.g. hollow fiber). The system is then pyrolyzed in an inert atmosphere and a dense membrane is obtained. An oxidation treatment is then necessary to create an open pore structure. Depending on the oxidation treatment typical molecules can be adsorbed and transported through the system.

Boutique (1986) has given an excellent description of pyrolization of PAN fibers (Figure 2.17). At low temperatures (200–300°C) and in an inert atmosphere, cyclization of PAN takes place (route A). At higher temperatures dehydrogenation takes place and a flat, dense structure of a cyclic carbon system develops. Under mild oxidation conditions (route B) the structure development is somewhat different and the intermediate structure is more open but the end structure is about the same (see also Fleming 1988). The fibers shrink strongly and become black. At higher temperatures graphitization takes place.

Koresh and Soffer (1980) used fibers or cloth (Carbone-Lorraine, France) for adsorption experiments with molecules like CO_2, O_2, Ar, N_2, C_2H_2 and H_2. The carbon samples were filled with "bonded" water at room temperature. On thermal treatment in vacuum at 700°C it was observed that the largest molecule which could be adsorbed was xenon. Treatment at higher

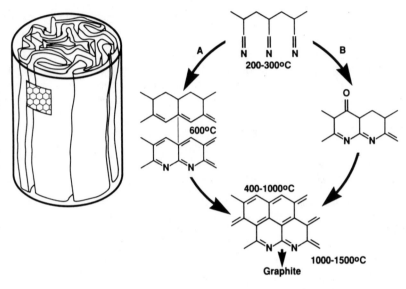

Figure 2.17. Structural model of carbon fibers and the transformation scheme of PAN to carbon during pyrolization (Fleming 1988).

Table 2.7. Some Modified Nanoscale Ceramic Microstructures Within Membranes With Pore Diameters of 3–5 nm (Burggraaf, Keizer and van Hassel 1989a, b)

Membrane Material	Modification by	Modified Structure	Size (nm)	Loading (wt%)
γ-Al$_2$O$_3$	Fe or V-oxide	Monolayer	≈ 0.3	5–10
γ-Al$_2$O$_3$	MgO/Mg (OH)$_2$	Particles		2–20
γ-Al$_2$O$_3$	Al$_2$O$_3$/Al (OH)$_3$	Particles		5–20
γ-Al$_2$O$_3$	Ag	Particles	5–20 > 20 nm	5–65
γ-Al$_2$O$_3$	CuCl/KCl	Multilayer		
γ-Al$_2$O$_3$	ZrO$_2$	Surface layer	< 1	2–25
γ-Al$_2$O$_3$	SiO$_2$ (amorphous)	20 nm layer + porous plugs	< 1.5	5–100
a-TiO$_2$*	V$_2$O$_5$	Monolayer	≈ 0.3	2–10
Al$_2$O$_3$/TiO$_2$	V$_2$O$_5$ or Ag	As for a-TiO$_2$ or Al$_2$O$_3$		
θ/α-Al$_2$O$_3$	ZrO$_2$/Y$_2$O$_3$	Multilayers/porous plugs	Few nm Pore size	1–100

* a-TiO$_2$: anatase titania

Table 2.8. Permeability, Selectivity and Separation Characteristics of Various Polymeric and Molecular Sieve Carbon (MSC) Membranes (Koresh and Soffer 1983)

Membrane	Gas	Permeability*	Selectivity[†]	Separation power
Cellulose acetate	He	0.136	97	13.2
	N_2	0.0014		
Polysulfone	H_2	0.12	40	4.8
	CO	0.003		
Silicon rubber	H_2	5.2	2.08	10.8
	CO	2.5		
	He	2.3	1.5	3.5
	N_2	1.5		
	O_2	3.96	2.15	8.3
	N_2	1.84		
Du Pont unspecified material	H_2	2.7–1.2	130–157	350–369
	N_2	0.02		
MSC membrane	O_2	17.1	7.1	124
	N_2	2.4		
	He	52	22	1140
	N_2	2.4		
	He	52	3.0	160
	O_2	17.1		
MSC-2	He	6	> 20	> 120
	O_2	< 0.3		
MSC-3	N_2	2.4*	> 24	> 58
	SF_6	< 0.1		

Separation power is the product of selectivity times permeability $\times 10^8$
* Unit of permeability is $cm^3 - cm^{-2} - cm - s^{-1} - cm\ Hg \times 10^{-8}$
[†] Selectivity is the ratio of single-gas permeabilities

temperatures showed pore closure due to sintering. The structure is opened up by thermal treatment in air (or oxygen) so that the larger molecules can adsorb. The relative affinity for adsorption in this type of systems was:

$$H_2O < CO_2 < C_2H_2 < H_2 < CO < N_2 (or\ Ar) < Xe < SF_6$$

The SF_6 molecules adsorb in this type of carbon system if the carbon fiber/cloth is oxidized at 400–450°C under mild oxidation conditions.

Koresh and Soffer (1983) developed a hollow-fiber gas separation membrane. In principle, polymeric hollow fibers can be porous (macroporosity) or dense. On thermal treatment in vacuum (pyrolysis) a second structural feature, the so-called ultramicroporosity (Koresh and Soffer 1983) was observed. This is due to small gaseous molecules channeling their way out of the solid matrix during pyrolysis. These micropores can be opened up by further activation (oxidation at moderate temperatures, 400–500°C) or closed by high-temperature sintering.

In gas separation applications, polymeric hollow fibers (diameter ≈ 100 μm) are used (e.g. PAN) with a dense skin. In the skin the micropores develop during pyrolyzation. This is also the case in the macroporous material but is not of great importance from gas permeability considerations. Depending on the pyrolysis temperature, the carbon membrane top layer (skin) may or may not be permeable for small molecules. Such a membrane system is activated by oxidation at temperatures of 400–450°C. The process parameters in this step determine the suitability of the asymmetric carbon membrane in a given application (Table 2.8).

2.7.2. Micro/Ultrafiltration Carbon Membranes

Le Carbone-Lorraine (France) produces asymmetric membranes for micro/ ultrafiltration in a composite carbon system (Fleming 1988) using a procedure similar to that described by Bird and Trimm (1983).

The porous carbon support is made by the pyrolysis of an 8 mm outer diameter tube with a wall thickness of 1 mm. Depending on the degree of pyrolysis (low temperature) the support can be weakly hydrophilic or more hydrophobic (high temperature). The latter material is weaker and more brittle.

On top of the support, thin (< 1 μm) polymeric films are deposited. For microfiltration, a second layer of phenolic resins in an organic solution is deposited, followed by controlled pyrolysis to the desired pore shape. To obtain smaller pores, a permselective carbon film can be produced, by adsorption of an appropriate monomer or oligomer, followed by in-situ polymerization. The degree of carbonization is controlled by "stepped" pyrolysis. In this manner, an inorganic carbon surface is formed which is very stable. High permeability can be achieved because of the large pore (> 10 μm) support. The membrane film is integrally bound to the support to allow backflushing in microfiltration applications.

The hollow-fiber systems for gas separation or the tubular microfiltration systems have to be pyrolyzed before mounting in the membrane housing, because of the large shrinkage during pyrolysis. That is the most critical step in the fabrication of a separation system.

2.7.3. Silica Membranes

In addition to polymers with C, H, O or N atoms, silicon rubbers also can be pyrolyzed as shown by Lee and Khong (1986). Silicon rubber tubes (Dow

Corning, diameter 1 mm, wall thickness about 0.2 mm) have been used to make pyrolyzed silicon-based membranes. Pyrolysis was carried out in a closed chamber in N_2 or He atmosphere at temperatures between 600 and 800°C followed by an oxidation step in air at temperatures of 500–900°C.

The remaining material after pyrolysis is believed to consist mainly of partially cross-linked Si–O chains. This structure becomes completely cross-linked in the oxidation step to form a refractory material of silica (SiO_2) as confirmed by chemical analysis of the final sample. The BET surface is of the order of 100 m^2/g and during the process a linear shrinkage of at least 10% was found which increased with increasing pyrolysis temperature. Weight loss was of the order of 25–60%. Although small cracks were observed in high magnification scanning electron microscope (SEM) gas permeability seemed to be of Knudsen diffusion type, which means that gas transport is independent of pressure and inversely proportional to the square root of molecular mass. The permeability of several gases such as He, H_2, O_2 and Ar varied between 0.7 and 7×10^{-4} cm^3 cm/cm^3-s-cm Hg. At the maximum porosity of 50%, the authors estimated a pore diameter of 5–10 nm. This suggests the formation of a mesoporous system contrary to the ultramicroporosity reported by Koresh, Jacob and Soffer (1983), who did not take into account the effect of cracks.

Membranes formed by pyrolysis such as carbon and others are amongst the most promising inorganic membranes for microfiltration/ultrafiltration as well as for gas separation under reducing or nonoxidizing environments. Polymeric tubular membrane structures can be pyrolized externally and then mounted in a membrane system. Fabricating and applying a membrane in one step (support and functional top layer) is possible as shown by Koresh, Jacob and Soffer (1986) but handling of these membranes is still difficult. A two-step (multistep) process (support preparation in the first step and top layer deposition in the second step) as shown by Fleming (1988) seems more flexible. The support preparation is relatively simple and it can be mounted easily in membrane systems. The functional carbon top layer can be treated by a different pyrolysis technique to obtain a structure with pores of 1 μm down to < 2 nm. The difficulty in this method can be the interaction between top layer and support during processing. In most cases it is necessary that the top layer thickness is smaller than 1 μm (sometimes 0.1 μm). With the multistep process, supports other than carbon can be used too.

2.8. TRACK-ETCH METHOD

An example of the track-etch membrane was given in Section 2.2 (Quinn et al. 1972). Booman and Delmastro (1974) have also described the layer deposition method to produce a microporous membrane by a track-etch method.

The track-etch membrane can be used in reverse osmosis and electrodialysis separation processes where it consists of a thin metal layer with a thin layer of insulator material on each side. The membrane pore diameters were in the range 0.5–10 nm. The membrane is made by depositing sequentially with a radio-frequency (rf) sputtering technique 0.5–50 nm thick layers on a 50–100 μm thick glass substrate to form a thin sandwich. The support glass substrate is chemically etched to produce 300 μm diameter holes by using a mask of sputter-etched molybdenum metal with a thin molybdenum layer functioning as a stopping layer for the etching step. Next the molybdenum masking layer is removed and, using the 300 μm holes in the substrate as a pattern, the molybdenum layer is chemically etched through. Next the resulting sandwich structure is irradiated with fission fragments to produce damage tracks in the Si_3N_4 layers, after which these layers are chemically etched along the fission damage tracks. Finally the metal membrane layer is chemically etched through, using the small holes in the surrounding Si_3N_4 dielectric layers as a mask.

2.9. COMPOSITE MEMBRANES: MODIFICATION METHODS

Membrane (top layer) structures are often able to achieve two specific objectives:

1. To obtain a further decrease of the effective pore size
2. To change the chemical nature of the internal surface

In both situations the interaction of the medium inside the pore with the pore wall (1) is increased (2) or changed which affect the transport and separation properties (surface diffusion, multilayer adsorption) and/or help overcome equilibrium constraints in membrane reactors. Membrane modifications can be performed by depositing material in the internal pore structure from liquids (impregnation, adsorption) or gases. Several modification possibilities are schematically shown in Figure 2.3. Some results obtained by Burggraaf, Keizer and coworkers are summarized in Table 2.7. Composite structures on a scale of 1–5 nm were obtained.

MgO- and Ag-modified membranes were obtained by homogeneous precipitation of the hydroxide from a typical solution consisting of 0.75 M urea and 0.2–0.5 M $AgNO_3$ or $Mg(NO_3)_2$ in water. The solution is introduced into the pores of the support and/or the γ-alumina top layer by impregnation. An increase in temperature results in (1) evaporation of the solvent and concentration of the solution and (2) the decomposition of urea (at $T > 90°C$) resulting in the formation of NH_3 and a decrease in the pH followed by precipitation of the metal hydroxide. The hydroxide is next converted to the oxide form at 350–450°C.

This modification method can be effected through several routes. With nonsupported membranes only relatively low loadings of 20 wt.% Ag or 2.5–3 wt.% MgO could be obtained even after repeated impregnation cycles. To increase the loading the solution must be concentrated within the pore system before precipitation. With supported membranes this can be performed by evaporation of the solvent through the top layer at 20–40°C at a relative humidity of 60%. This results in the concentration of the metal salt from the support into the top layer. Subsequently the temperature is increased to the point where the urea decomposes. In the case of the $AgNO_3$–urea system the reaction rate is high and concentration of metal salt in the top layer was found to be the only way to avoid precipitation in the support (van Praag et al. 1990). After calcination in H_2 atmosphere a crack-free membrane with a maximum loading of 65 wt.% Ag (based on the top layer) could be obtained (this is about 50% of the pore volume). The crystallite size was about 5–20 nm. With $Mg(NO_3)_2$–urea the reaction rate was found to be much slower and evaporation and reaction can be performed simultaneously at 90°C, resulting in 20 wt.% MgO loading. This MgO is homogeneously distributed across the thickness of the top layer. Scanning auger microscopy (SAM), XRD and TEM investigations showed that all the metal is indeed concentrated in the crack-free top layer. The precise distribution is currently under investigation and will be reported in future. Modeling considerations showed that (1) the concentration profile can be influenced by changing the ratio of evaporation/concentration and reaction rate and (2) for γ-alumina membranes a decrease of the effective pore size can be expected only when the maximum loading exceeds the level of about 40 vol.% of the pore volume. At lower levels, indeed no decrease of the effective pore size (from the desorption branch) was found. Vanadia-modified membranes were synthesized (van Praag et al. 1990) by impregnation of the γ-alumina membrane with a vanadium acetylacetonate solution to obtain monolayers of vanadium oxide (van Ommen et al. 1983). The vanadium acetylacetonate reacts with the OH group of the membrane material to form a surface group //|–O–V–O (AcAc). After calcination at 450°C a surface vanadium oxide group is formed which is structurally related to V_2O_5. The chemical reactivity of the surface is completely changed as indicated by the inactivity of the methanol dehydrogenation reaction of the modified material unlike that of the nonmodified material. This indicates a homogeneous coverage of the alumina with vanadia. Very interesting results were obtained by the modification of γ-alumina top layers with silica (Uhlhorn et al. 1989). Here first a silica polymeric gel solution was prepared by hydrolysis of a TEOS (tetraethyl orthosilicate) solution in ethanol with HNO_3 as a catalyst. A γ-alumina membrane with pores of 2.5–3 nm was brought in contact with a diluted dispersion of the polymeric gel solution used as the dipping medium. After a short drying period the dipped membrane was calcined at 600°C. The shape of the adsorption–desorption isotherm indicated that the effective pore size was decreased to below one nanometer radius. Preliminary SAM results indicate the presence of a silica layer of about 20 nm and of silica "props" penetrating the pores in the alumina. The permselectivity and the separation factor of a number of gas mixtures involving CO_2, propane, propene, butane etc. with inert gases is strongly increased (up to 30) which support the existence of very small pores. Similar results are reported by Asaeda and Du (1986). These authors treated a γ-alumina membrane with Al-isopropoxide dissolved in an organic solvent followed by treatment with sodium silicate solution in water. Finally the heated membrane was kept in humid air at about 90°C for a few days. At temperatures lower than 90°C, the membranes thus produced showed a high separation factor (60–100) for alcohol/water mixtures as compared with the nonmodified ones (factor 5–10). From these results a pore size of about 1 nm was estimated. Aluminosilicate hydroxide "props" are probably present which collapse at temperatures higher than 90°C. This behavior is in contrast with the silica-modified membranes synthesized by Burggraaf, Keizer and van Hassel (1989a, b) where the separation properties remained unchanged even after heat treatment at high temperature.

Recently the synthesis of zeolitic membranes was reported by Suzuki (1987). Initially, an Al_2O_3 (or silica) layer is formed by coating a porous alumina or Vycor glass support with an Al-acetate solution in aqueous acetic acid followed by calcining the acetate at 500°C. The oxide layer formed in this way is then transformed into a gel layer by heat treatment over an 18 h period at 40°C in a KOH/NaOH solution at pH 10. The adjacent surface layer of the support is also gelled by this procedure. Subsequently the system was hydrothermally heated at 100°C for 4 days in an autoclave. According to angle-resolved XPS measurements a 2–10 nm thick layer of an "NaA"-type zeolite (sodalite) film is formed by in-situ crystallization at the entrance of the 4 nm pores of the Vycor glass support. A further modification of pores in the zeolite range seems possible according to results of Niwa et al. (1984). A pore size reduction to 0.1 to 0.2 nm by chemical vapor deposition (CVD) was reported due to the formation of 2–4 layers of SiO_2 from $Si(OCH_3)_4$ on top of the mordenite crystal.

2.10. MISCELLANEOUS METHODS AND COMMENTS

Besides the synthesis methods for porous membranes and their modification methods discussed above, other synthesis methods have been reported. These are outlined below. Preparation of dense membranes is discussed in Section 2.2. The other types are the so-called dynamically formed membranes which consist of hydroxides or salts formed in the pores of a support during the separation process (Woerman 1978, Freilich and Tanny 1978). For further discussion of these types of membranes refer to Chapter 4. Physical methods like sputtering, e.g. of alumina, silica or metal layers can result in interesting effects. These are, however, of limited practical use in view of the small surface area structures obtainable by such techniques and are therefore not discussed here.

An important aspect of all (ceramic) membranes is that defects (pinholes, cracks) present in appreciable amounts drastically affect the separation properties. Membranes should therefore be defect-free. Surprisingly very little has been reported on the characterization of membrane structures. In many publications none or only visual inspection is used to characterize defects. The most sensitive tool for detecting defects however is the gas permeability measurement. A method for characterization of the presence of defects based on gas permeability was developed by Vuren et al. (1987). The observed permeability values can be due to the contributions of the support and of the top layer. Defect-free top layers must have a permeability independent of the average pressure (Knudsen diffusion only). Defects of appreciable size causes a laminar flow contribution (pressure-dependent behavior); the same holds for the supported system. The absence of pressure dependency of the gas permeability of a nonadsorbing gas through the top layer is therefore considered as an indication that no defects are present. This technique has been very successfully applied by Burggraaf, Keizer and van Hassel in the synthesis of a variety of membrane structures. For example zirconia and titania membranes reported by Cot, Guizard and Larbot (1988) in Section 2.4 are probably not crack-free because the calculation of their liquid (and gas) permeability based on the reported microstructural data yields much smaller fluxes and much stronger flow dependency on pore size.

REFERENCES

Anderson, M. A., M. L. Gieselman, and Q. Xu. 1988. Titania and alumina ceramic membranes. *J. Membrane Sci.* 39: 243–58

Anotec Separations. 1986. Anopore—A new inorganic membrane unique in the world of microfiltration. Brochure.

Asaeda, M. and L. D. Du. 1986. Separation of alcohol/water gaseous mixtures by thin ceramic membrane. *J. Chem. Eng. Japan* 19(1): 72–77, 84–85.

Auriol, A. and D. Tritten. 1973. A process for the manufacture of porous filter supports. French Patent 2,463,636.

Baeyer, H. V., R. Schnabel, W. Vaulont and P. Kaczmarczyk. 1980. Properties of porous glass membranes with respect to applications in blood purification. *Trans. Am. Soc. Artif. Intern. Organs* 26: 309–13.

Bansal, I. K. 1975. Ultrafiltration of oily wastes from process industries. *A.I.Ch.E. Symp. Ser.* 71(151): 93–99.

Bansal, I. K. 1976. Concentration of oily and latex waste waters using ultrafiltration inorganic membranes. *Ind. Water Engr.* 13(5): 6–11.

Bird, A. J. and D. L. Trimm. 1983. Carbon molecular sieves used in gas separation membranes. *Carbon* 21(3): 177–80.

Booman, G. L. and R. J. Delmastro. 1974. Porous metal insulator sandwich membrane. U.S. Patent 3,794,174.

Boutique, J.-P. 1986. Carbon fibres: Preparation, performance and utilizations. *Rev. M. Mec.* 29(3–4): 183–89.

Brinker, C. J., D. E. Clark and D. R. Ulrich, eds. 1984. *Better Ceramics Through Chemistry.* Mater. Res. Soc. Symp. Ser. 32. Elsevier, New York.

Brinker, C. J., D. E. Clark and D. R. Ulrich, eds. 1988. *Better Ceramics Through Chemistry.* Mater. Res. Soc. Symp. Ser. 121. Pittsburgh.

Burggraaf, A. J., K. Keizer and B. A. van Hassel. 1989a. Nanophase ceramics, membranes and ion implanted layers. Paper read at S.I.C. Mat. 88-Nato ASI, Surfaces and interfaces of ceramic materials, 4–16 September 1988, Ile d'Oléron.

Burggraaf, A. J., K. Keizer and B. A. van Hassel. 1989b. Ceramic nanostructure materials, membranes and composite layers. *Solid State Ionics* 32/33 (Part 2): 771–82.

Cacciola, A. and P. S. Leung. 1981. Process for the production of a dry inorganic ultrafiltration membrane and membrane produced by such a method. European Patent Appl. 040,282.

Cot, L., C. Guizard and A. Larbot. 1988. Novel ceramic material for liquid separation process: Present and prospective applications in microfiltration and ultrafiltration. *Industrial Ceramics* 8(3): 143–48.

Deryaquin, B. V. and S. M. Levi. 1964. *The Physical Chemistry of Coating Thin Films on a Moving Support.* The Focal Press. London.

Dunbobbin, B. R. and W. R. Brown. 1987. Air separation by a high temperature molten salt process. *Gas separation and purification* 1: 23–29.

Eguchi, K., H. Tanaka, T. Yazawa and T. Yamagura. 1987. Chemically durable porous glass and process for the manufacture thereof. European Patent Appl. 0,220,764.

Elmer, T. H. 1978. Evaluation of porous glass as desalination membrane. *Ceramic Bull.* 57(11): 1051–53, 60.

Fleming, H. L. 1988. Carbon composites: a new family of inorganic membranes. Paper read at 6th Annual Membrane Planning Conference, 1 November 1988, Cambridge. MA.

Freilich, D. and G. B. Tanny. 1978. The formation mechanism of dynamic hydrous Zr (IV) oxide membranes on microporous supports. *J. Coll. Interface Sci.* 64(2): 362–70.

Furneaux, R. C., W. R. Rigby and A. P. Davidson. 1985. Porous films and method of forming them. European Patent Appl. 0,178,831.

Furneaux, R. C., W. R. Rigby and A. P. Davidson. 1989. The formation of controlled-porosity membranes from anodically oxidized aluminum. *Nature* 337: 147–49.

Furneaux, R. C., G. E. Thompson and G. C. Wood. 1978. The application of ultramicrotomy to the electronoptical examination of surface films on aluminum. *Corrosion Science* 18: 853–81.

Gallagher, D. and L. C. Klein. 1986. Silica membranes by the sol–gel process. *J. Coll. Interface Sci.* 109(1): 40–45.

Gerster, D. and R. Veyre. 1985. Mineral ultrafiltration membranes in industry. In ACS Symp. Ser. 281 *Reverse Osmosis and Ultrafiltration*, eds. S. Sourirajan and T. Matsuura, pp. 225–30.

Gieselman, M. J., M. A. Anderson, M. D. Moosemiller and C. G. Hill. 1988. Physico-chemical properties of supported and unsupported τ-Al$_2$O$_3$ and TiO$_2$ ceramic membranes. *Separation Science and Technology* 23: 1695–714.

Gillot, J. 1987. Filtration membrane and preparation process of such a membrane. European Patent Appl. 0,092,840 Bl.

Grangeon, A. 1987. Monolithic ultrafiltration or microfiltration module made of carbon or porous graphite and its manufacturing process. French Patent 2,585,965.

Guizard, C., N. Cygankewiecz, A. Larbot and L. Cot. 1986a. Sol–gel transition in zirconia systems using physical and chemical processes. *J. Non-Cryst. Solids* 82: 86–92.

Guizard, C., N. Idrissi, A. Larbot and L. Cot. 1986b. An electronic conductive membrane from a sol–gel process. *Proc. Brit. Ceramic Soc.* 38: 263–75.

Guizard, C., F. Legault, N. Idrissi, L. Cot and C. Gavach. 1989. Electronically conductive mineral membranes designed for electroultrafiltration. *J. Membrane Science* 41: 127–42.

Hammel, J. J., W. P. Marshall, W. J. Robertson and H. W. Barch. 1987. Porous siliceous-containing gas enriching material and process of manufacture and use. European Patent Appl. 0,248,931.

Haraya, K., Y. Shindo, T. Hakuta and H. Yoshitome. 1986. Separation of H_2-CO mixtures with porous glass membranes in the intermediate flow region. *J. Chem. Soc. Japan* 19(3): 186–190.

Hench, L. L. and D. R. Ulrich, eds. 1986. *Science of Ceramic Chemical Processing*. Wiley Interscience Publications, New York.

Hench, L. L. and D. R. Ulrich, eds. 1984. *Ultrastructure Processing of Ceramics, Glasses and Composites*. Wiley Interscience Publications, New York.

Hoar, T. P. and N. F. Mott. 1959. Mechanism for the formation of porous anodic oxide films on aluminum. *J. Phys. Chem. Solids* 9: 97–99.

Hood, H. P. and M. E. Nordberg. 1938. Heat treated article of borosilicate glass. U.S. Patent 2,106,744.

Hsieh, H. P. 1988. Inorganic membranes. *A.I.Ch.E. Symp. Ser.* 261. (84): 1–18.

Iler, R. K. 1979. *The Chemistry of Silica*. John Wiley & Sons, New York.

Itaya, K., S. Sugawara, K. Arai and S. Saito. 1984. Properties of porous anodic aluminum oxide films as membranes. *J. Chem. Eng. Japan* 17(5): 514–20.

Itoh, N., Y. Shindo, K. Haraya and T. Hakuta. 1988. A membrane reactor using microporous glass for shifting equilibrium of cyclohexane dehydrogenation. *J. Chem. Eng. Japan* 21(4): 399–404.

Kaiser, A. and H. Schmidt. 1984. Generation of silica-membranes from alkoxisilanes on porous supports. *J. Non-Cryst. Solids* 63: 261–71.

Keizer, K. and A. J. Burggraaf. 1988. Porous ceramic materials in membrane applications. *Science of Ceramics*, ed. D. Taylor, 14: pp. 83–95.

Klein, L. C. and D. Gallagher. 1988. Pore structures of sol–gel silica membranes. *J. Membrane Science* 39(3): 213–21.

Koresh, J. and A. Soffer. 1980. Study of molecular sieve carbons. Part 1—Pore structure, gradual pore opening and mechanism of molecular sieving. *J.C.S. Faraday* I 76: 2457–71.

Koresh, J. E. and A. Soffer. 1986. Mechanism of permeation through molecular-sieve carbon. *J.C.S. Faraday* I 82: 2057–63.

Koresh, J. E. and A. Soffer. 1987. The carbon molecular sieve membranes. General properties and the permeability of CH_4/H_2 mixtures. *Separation Science and Technology* 22(2&3): 972–82.

Koresh, J. E. and A. Soffer. 1983. Molecular sieve carbon permselective membrane. Part I. Presentation of a new device for gas mixture separation. *Separation Science and Technology* 18(8): 723–34.

Kraus, K. A., A. E. Marcinkowsky, J. S. Johnson and A. J. Shor. 1966. Salt rejection by a porous glass. *Science* 151: 194–5.

Larbot, A., J. A. Alary, C. Guizard, L. Cot and J. Gillot. 1987. New inorganic ultrafiltration membranes: Preparation and characterization. *Int. J. High Technology Ceramics* 3: 145–51.

Larbot, A., J. P. Fabre, C. Guizard and L. Cot. 1988. Inorganic membranes obtained by sol–gel techniques. *J. Membrane Science* 39(3): 203–13.

Larbot, A., A. Julbe, C. Guizard and L. Cot. 1989. Silica membranes by the sol–gel process. *J. Membrane Science* 44: 289–303.

Larbot, A., J.-P. Fabre, C. Guizard and L. Cot. 1989. New inorganic ultrafiltration membranes: titania and zirconia membranes. *J. Amer. Ceram. Soc.* 72(2): 257–61.

Lee, K. H. and S. J. Khang. 1986. A new silicon-based material formed by pyrolysis of silicone rubber and its properties as a membrane. *Chem. Eng. Commun.* 44(1–6): 121–32.

Leenaars, A. F. M. and A. J. Burggraaf. 1985. The preparation and characterization of alumina membranes with ultrafine pores: 2. The formation of supported membranes. *J. Coll. Interface Sci.* 105(1): 27–40.

Leenaars, A. F. M., K. Keizer and A. J. Burggraaf. 1984. The preparation and characterization of alumina membranes with ultra-fine pores. Part 1. Microstructural investigations on non-supported membranes. *J. Mater. Science* 19: 1077–88.

Leenaars, A. F. M., A. J. Burggraaf and K. Keizer. 1987. Process for the production of crack-free semi-permeable inorganic membranes. U.S. Patent. 4,711,719.

Livage, J. 1986. The gel route to transition metal oxides. *J. Solid State Chem.* 64: 322–30.

Livage, J. and M. Henry. 1985. A predictive model for inorganic polymerization reactions. In *Ultrafiltration Processing of Advanced Ceramics.* eds. J. D. Mackenzie and D. R. Ulrich, pp. 183–95. John Wiley & Sons, New York.

McMillan, P. W. 1980. Tomorrow's techniques; Latest glass technology improves efficiency and range of membrane processes. *Process Engineering* April: 41.

McMillan, P. W. and R. Maddison. 1981. Improvements in or relating to porous glass. European Patent Appl. 0,039,179.

McMillan, P. W. and C. E. Matthews. 1976. Microporous glass for reverse osmosis. *J. Mater. Science* 11: 1187–99.

Messing, G., E. Fuller and H. Hausner, eds. 1988. *Ceramic Powder Processing.* Westerville, OH. Am. Ceram. Soc. Inc.

Messing, J. 1978. Hydrophobic inorganic membrane. German Patent DE 2,905,353.

Mitrovic, M. and L. Knezic. 1979. Electrolytic aluminum oxide membranes—a new kind of membrane with reverse osmosis characteristics. *Desalination* 28: 147–56.

Niwa, M., S. Kato, T. Hattori and Y. Murikami. 1984. Fine control of the pore opening size of zeolite mordenite by chemical vapor deposition of silicon alkoxide. *J.C.S. Faraday Trans.* I 80: 3135–45.

O'Sullivan, J. P. and G. C. Wood. 1970. The morphology and mechanism of formation of porous anodic films on aluminum. *Proc. Roy. Soc. London* A317: 511–43.

Ohya, H., Y. Tanaka, M. Niwa, R. Hongladaromp, Y. Negismi and K. Matsumoto. 1986. Preparation of composite microporous glass membrane on ceramic tubing. *Maku* 11: 41–44.

Ommen, J. G. van, K. Hoving, H. Bosch, A. J. van Hengstum and P. J. Gellings. 1983. The preparation of supported oxide catalysts by adsorption of metal acetylacetonates, M(AcAc)n on different supports. *Z. Phys. Chemie Neue Folge* 134: 99–106.

Partlow, D. E. and B. E. Yoldas. 1981. Colloidal versus polymer gels and monolithic transformation in glass forming systems. *J. Non-Crystalline Solids* 46: 153–161.

Pavlovic, T. and A. Ignatiev. 1986. Optical and microstructural properties of anodically oxidized aluminum. *Thin Solid Films* 138: 97–109.

Pez, G. P. 1986. Method for gas separation. U.S. Patent 4,612,209.

van Praag, W., V. Zaspalis, K. Keizer, J. G. van Ommen, J. R. Ross and A. J. Burggraaf. 1990. Preparation, modification and microporous structure of alumina and titania ceramic membrane systems. Proc. *1st. Intl. Conf. Inorganic Membranes*, 3–6 July 1989, 397–400, Montpellier.

Quinn, J. A., J. L. Anderson, W. S. Ho and W. J. Petzny. 1972. Model pores of molecular dimension. Preparation and characterization of track-etched membranes. *Biophys. J.* 12(8): 990–1007.

Ramsay, S. R., S. R. Daish and C. J. Wright. 1978. Structure and stability of concentrated boehmite sols. *Farad. Disc. Chem. Soc.* 65: 66–74.

Scherer, G. W. 1986. Drying gels I. General theory. *J. Non-Crystalline Solids* 87: 199–225.

Scherer, G. W. 1987. Drying gels II. Film and flat plate. *J. Non-Crystalline Solids* 89: 217–38.

Schnabel, R. and W. Vaulont. 1978. High pressure techniques with porous glass membranes. *Desalination* 24: 249–72.

Schnabel, R. 1976. Separation membranes from porous glass and method to produce them. German Patent 2,454,111.

Schnabel, R., P. Langer and S. Breitenbach. 1988. Separation of protein mixtures by BIORAN porous glass membranes. *J. Membr. Science* 36: 55–66.

Sing, K. S. W., D. H. Everett, R. A. W. Haul, L. Moscou, R. A. Pierotti, J. Rouquerol and T. Siemieniewska. 1985. Reporting physisorption data for gas/solid systems with special reference to the determination of surface area and porosity. *Pure and Appl. Chem.* 57(4): 603–19.

Smith, A. W. 1973. Porous anodic aluminum oxide membrane. *J. Electrochem. Soc.* 120(8): 1068–69.

Smith, A. W. 1974. Process for producing an anodic aluminum oxide membrane. U.S. Patent 3,850,762.

Suzuki, H. 1987. Composite membrane having a surface layer of an ultrathin film of cage-shaped zeolite and processes for production thereof. U.S. Patent 4,699,892.

Takahashi, H., M. Nagayama, H. Akahori and A. Kitahara. 1973. Electron-microscopy of porous anodic films on aluminum by ultrathin section technique. Part 1. The structural change of the film during the current recovery. *J. Electron Microscopy* 22(2): 149–57.

Tallmadge, J. A. and C. Gutfinger. 1967. Entrainment of liquid films, drainage withdrawal and removal. *Ind. Eng. Chem.* 59(11): 19–34.

Thompson, G. E., R. C. Furneaux, G. C. Wood, J. A. Richardson, and J. S. Goode. 1978. Nucleation and growth of porous anodic films on aluminum. *Nature* 272: 433–35.

Terpstra, R. A., B. C. Bonekamp and H. J. Veringa. 1988. Preparation, characterization and some properties of tubular alpha alumina ceramic membranes for microfiltration and as a support for ultrafiltration and gas separation membranes. *Desalination* 70: 395–404.

Tiller, F. M. and Chum-Dar Tsai. 1986. Theory of filtration of ceramics: I, Slip casting. *J. Amer. Ceram. Soc.* 69(12): 882–87.

Trimm, D. L. and B. J. Cooper. 1970. The preparation of selective carbon molecular sieve catalysts. *Chem. Commun.* No. 8: 477–78.

Trimm, D. L. and B. J. Cooper. 1973. Propylene hydrogenation over platinum/carbon molecular sieve catalysts. *J. Cat.* 31: 287–92.

Uhlhorn, R. J. R., M. H. B. J. Huis in't Veld, K. Keizer and A. J. Burggraaf. 1988. New ceramic membrane materials for use in gas separation applications. *Science of Ceramics* 14: 551–56.

Uhlhorn, R. J. R., M. H. B. J. Huis in't Veld, K. Keizer and A. J. Burggraaf. 1989. High permselectivities of microporous silica-modified τ-alumina membranes. *J. Mater. Science Lett.* 8(10): 1135–39.

Uhlhorn, R. J. R., K. Keizer and A. J. Burggraaf. 1989. Formation of and gas transport properties in ceramic membranes. In: *Advances in Reverse Osmosis and Ultrafiltration*, eds. T. Matsuura and S. Sourirajan, pp. 239–59. Nat. Res. Council Canada, Ottawa.

Veyre, R. 1985a. Ultrafiltration mineral membranes for liquid sterilization. *Inf. Chim.* 261: 187–91.

Veyre, R. 1985b. Industrial membranes and perspective on development of the 3rd generation Carbosep® ultrafiltration membranes. *Ann. Chim. Fr.* 10: 359–72.

Vuren, R. J. V., B. C. Bonekamp, K. Keizer, K., R. J. R. Uhlhorn, H. Veringa and A. J. Burggraaf. 1987. Formation of ceramic alumina membranes for gas separation. In *Materials Science Monographs*, ed. P. Vincenzini, 38c: pp. 2235–45.

Woermann, D. 1978. Inorganic precipitation membranes. *Naturwissenschaften* 74: 528–35.

Yoldas, B. E. 1975. Alumina gels that form porous transparent Al_2O_3. *J. Mater. Science* 10: 1856–60.

Zaspalis, V. T., K. Keizer, W. van Praag, J. G. van Ommen, J. G., J. R. H. Ross and A. J. Burggraaf. 1989. Ceramic membranes as catalytic active materials in selective (oxidative) dehydrogenation processes. *Proc. Brit. Ceramic Soc.* 43: 103–39.

3. General Characteristics of Inorganic Membranes

H. P. HSIEH*

Alcoa Laboratories, Alcoa Center, PA

3.1. INTRODUCTION

The separation efficiency (e.g. permselectivity and permeability) of inorganic membranes depends, to a large extent, on the microstructural features of the membrane/support composites such as pore size and its distribution, pore shape, porosity and tortuosity. The microstructures (as a result of the various preparation methods and the processing conditions discussed in Chapter 2) and the membrane/support geometry will be described in some detail, particularly for commercial inorganic membranes. Other material-related membrane properties will be taken into consideration for specific separation applications. For example, the issues of chemical resistance and surface interaction of the membrane material and the physical nature of the module packing materials in relation to the membranes will be addressed.

3.2. COMMERCIAL INORGANIC MEMBRANES

Inorganic membranes commercially available today are dominated by porous membranes, particularly porous ceramic membranes which are essentially the side-products of the earlier technical developments in gaseous diffusion for separating uranium isotopes in the U.S. and France. Summarized in Table 3.1 are the porous inorganic membranes presently available in the market (Hsieh 1988). They vary greatly in pore size, support material and module geometry.

To help understand the performance of membranes, brief explanations of a few terminologies are in order. Permeability of a membrane is determined by dividing permeate flux by the transmembrane pressure. It indicates the membrane's throughput per unit area (flux) per unit pressure difference. An important factor affecting flux and retention ability of the membrane is the direction of the feed flow relative to the membrane surface. In through-flow configuration, the feed flow is perpendicular to the membrane surface. In cross-flow configuration, the feed stream flows parallel to the membrane

---

* With R. R. Bhave.

Table 3.1. Commercial Porous Inorganic Membranes

Manufacturer	Trade Name	Membrane Material	Support Material	Membrane Pore Diameter	Geometry of Membrane Element	Tube or Channel Inside Diameter (mm)
Alcoa/SCT	Membralox®	ZrO_2 Al_2O_3	Al_2O_3 Al_2O_3	20–100 nm 0.2–5 μm	Monolith/ Tube	4 and 6
Norton	Ceraflo®	Al_2O_3	Al_2O_3	0.2–1.0 μm 6 μm (symmetric)	Monolith Tube	3
NGK		Al_2O_3	Al_2O_3	0.2–5 μm	Tube	7 and 22
Du Pont	PRD-86	Al_2O_3, Mullite, Cordierite	None	0.06–1 μm	Tube	0.5–2.0
Alcan/Anotec	Anopore®	Al_2O_3 Al_2O_3	Al_2O_3 Al_2O_3	20 nm 0.1 μm 0.2 μm	Plate	
Gaston County Filtration Systems	Ucarsep®	ZrO_2	C	4 nm	Tube	6
Rhone-Poulenc/SFEC	Carbosep®	ZrO_2 ZrO_2	C C	∼ 4 nm 0.08–0.14 μm	Tube	6
Du Pont/ CARRE		$Zr(OH)_4$	SS	0.2–0.5 μm	Tube	∼ 2
TDK	Dynaceram®	ZrO_2	Al_2O_3	∼ 10 nm	Tube	⩽5
Asahi Glass		Glass	None	8 nm–10 μm	Tube/Plate	3 and 10
Schott Glass		Glass	None	10 nm and 0.1 μm	Tube	5–15
Fuji Filters		Glass Glass	None None	4–90 nm 0.25–1.2 μm	Tube	
Ceram-Filtre	FITAMM	SiC	None	0.1–8 μm	Monolith	25
Fairey	Strata-Pore® Microfiltrex®	Ceramics SS	Ceramics SS	1–10 μm 0.2–1 mm	Tube/Plate Tube Plate	10
Mott		SS, Ni, Au, Ag, Pt, etc.	None	⩾ 0.5 μm	Tube	3.2–19
Pall		SS, Ni, etc.	None	⩾ 0.5 μm	Tube	60 and 64
Osmonics	Hytrex® Ceratrex®	Ag Ceramics	None Ceramics	0.2–5 μm 0.1 μm	Tube/Plate	
Ceramem		Ceramics oxides	Coerdierite	0.05–0.5 μm	Honeycomb monolith	1.8

surface and the cross-flow velocity can have a significant influence on flux, as well as retention characteristics. These aspects are treated in detail in Chapter 5.

Among the commercially available inorganic membranes, zirconia- and alumina-based membranes are used in large-scale applications in a wide variety of areas.

Many zirconia membrane systems come in the multiple tubular configuration and may contain more than a thousand tubes in a module (Gaston County Filtration Systems 1980). Their support materials are either carbon, stainless steel or alumina. The membranes are claimed to be usable over a pH range 0–14 although the preferred operating range is narrower. Water permeability for Ucarsep® -type zirconia membranes with a pore diameter in the 4–20 nm range has been given as approximately 45–250 L/h-m²-bar at ambient temperatures under a recommended cross-flow velocity 2.5–7 m/s (Bansal 1976, Gerster and Veyre 1985). For more open zirconia membranes (Membralox® from Alcoa/SCT) with pore diameters 50–100 nm, the water permeability is approximately 800–1500 L/h-m²-bar (Gillot, Soria and Garcera 1990). Although the membrane itself can be used at high temperatures, the systems which include gaskets (potting compounds) and end-seal materials are generally recommended for operation below 150°C.

Commercial alumina membranes are the extension of technological developments on gas diffusion in the nuclear industry. Since their introduction by Alcoa/SCT and Norton in the early 1980s, alumina membranes have been used in many applications. Hsieh, Bhave and Fleming (1988) have given a comprehensive review of alumina membranes from preparation methods to applications and operating guidelines. The microfiltration alumina membranes which are made of α-alumina can withstand a very high temperature (> 1000°C). The systems which contain other auxiliary materials such as gaskets and end seals, however, are recommended for temperatures up to 130°C when steam sterilization is conducted (Gillot and Garcera 1984). A cross-flow velocity of 5 m/s is recommended for fouling control. The water permeability is approximately 2000 L/h-m²-bar for 0.2 micron membranes. The ultrafiltration alumina membranes have a much smaller permeability, for example, 10 L/h-m²-bar for 4 nm membranes (Alcoa 1987) and are made of transition-phase alumina. They do not have as good resistance to strong acids or strong bases as the α-alumina.

Another class of alumina membranes made by an entirely different process (anodic oxidation) are marketed by Alcan/Anotec and used exclusively for laboratory applications as these are currently available only in disk forms.

Porous glass structures have not been marketed for membrane separation applications until recently, despite having been studied as a membrane material for a long time. A number of glass companies such as Asahi Glass and Schott Glass have introduced porous glass in the form of capillary bundles, tubes and plates for membrane liquid separations as in medical and biotechnology applications. Certain properties of glass membranes are similar to those of alumina and zirconia membranes described earlier (see also Section 2.5). Both types of membranes are steam sterilizable, mechanically and chemically stable in the process of cleaning, biologically inert and resistant to attack by microorganisms. It should be noted, however, that silica dissolution from glass membranes in aqueous solutions has been known to occur to a varying degree and some chemical treatments are available for retarding that dissolution process. Such treatments include the addition of aluminum chloride in the feed stream (Ballou, Leban and Wydeven 1973) and surface modification of the membranes (Schnabel and Vaulont 1978). Glass membranes generally have a maximum working temperature of 700–800°C (Kameyama et al. 1981). Their water permeability is reported to be in the range 1–1.5 L/h-m^2-bar for ultrafiltration membranes having a pore diameter of 3–4 nm. As expected, the water permeability of reverse osmosis type glass membranes is significantly lower in the range 0.01–0.1 L/h-m^2-bar (Schnabel and Vaulont 1978).

Porous metals have long been commercially available for particulate filtration. They have been used in some cases as microfiltration membranes that can withstand harsh environments, or as porous supports for dynamic membranes. Stainless steel is by far the most widely used porous metal membrane. Other materials include silver, nickel, Monel, Hastelloy and Inconel. Their recommended maximum operating temperatures range from 200 to 650°C. Depending on the pore diameter which varies from 0.2 to 5 microns, the water permeability of these symmetric membranes can exceed 3000 L/h-m^2-bar and is similar to that obtained with asymmetric ceramic microfiltration membranes. Due to the relatively high costs of these membranes, their use for microfiltration has not been widespread.

3.3. MICROSTRUCTURAL CHARACTERISTICS

3.3.1. Microscopic Morphology

A scanning electron microscope (SEM) generates electron beams and forms an image from the emitted electrons as a result of interaction between the bombarding electrons and the atoms of the specimen. Since electrons have a much shorter wavelength than light photons, SEMs can generate higher-resolution information than reflected light microscopes. With their improved resolution and competitive pricing, SEMs have become a basic surface and microstructural characterization tool in membrane separations. This is particularly true for porous materials such as porous inorganic membranes where the three-dimensional appearance of textured surfaces can be revealed by the depth-of-field feature of a SEM. Theoretically, the maximum magnification of electron beam instruments such as SEMs can be beyond 800,000 ×. Practical magnification and resolution limits, however, are less than 100,000 × (and often less than 30,000 × for less expensive models) due to instrumental parameters.

Some commercial inorganic membranes such as porous glass and metal membranes have a symmetric or homogeneous microstructure (Figure 3.1). However, the majority of the commercially important inorganic membranes are asymmetric and composite in nature. They usually consist of a thin fine-pore film responsible for separating components, and a support or substrate with single or multiple layers having larger pores for imparting the required mechanical strength to the membrane composite. Figure 3.2 shows the scanning electron micrographs of the cross sections of membrane composites with a single-layer support (Figure 3.2a) and with a multiple-layer support (Figure 3.2b).

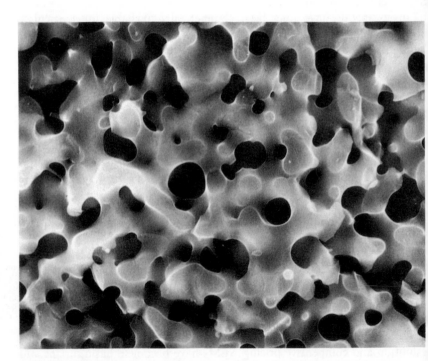

Figure 3.1. Scanning electron micrograph of the cross-section of a porous glass membrane with symmetric microstructure (courtesy of Asahi Glass).

The basic idea behind the composite and asymmetric structure is to minimize the overall hydraulic resistance of the permeate flow path through the membrane structure. The permeate flux through a given layer is inversely proportional to the layer thickness and is, under simplified assumptions, proportional to some power of the pore size of the porous layer. It is, therefore, desirable to have a separative layer (membrane) as thin as possible and yet possessing defect-free physical integrity, and one or more layers of supports which provide the necessary mechanical strength with negligible hydraulic resistance. This also helps reduce the pressure required for back-flushing in microfiltration operation. In cases where the precursor particles of the membrane layer are very small in size compared to the pore size of the bulk of the support, the membrane particles can significantly penetrate the support pores and the resulting permeability of the membrane/support composite will deteriorate. A practical solution is to add one or more intermediate layers having pore sizes between those of the membrane layer and the bulk support.

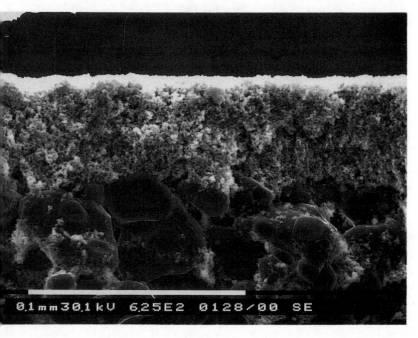

(a)

Figure 3.2. Scanning electron micrographs of the cross-sections of alumina membrane composites with (a) one layer of support and (b) three layers of support (Hsieh, Bhave and Fleming 1988).

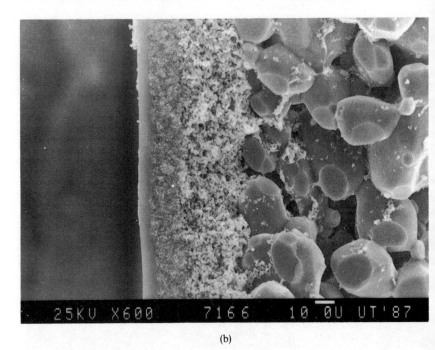

(b)

Figure 3.2 (*continued*)

For efficient separation, porous inorganic membranes need to be crack-free and uniform in pore size. An important reason for the increasing acceptance of ceramic membranes introduced in recent years is the consistent quality as exemplified in a scanning electron micrograph of the surface of a 0.2 micron pore diameter alumina membrane (Figure 3.3).

SEM analysis can be coupled with a spectrometer capable of detecting X-rays emitted by the specimen during electron-beam excitation. These X-rays carry a unique energy level and wavelength which can reveal the elemental composition of the specimen. This type of X-ray microanalysis commonly used in conjunction with SEMs is known as energy-dispersive X-ray analysis (EDX or EDXRA). It can identify a large number of elements (except light elements such as carbon, nitrogen and oxygen) on a small area (practically down to a fraction of a micron, depending on the material to be analyzed). This technique can be useful, for example, in studying the nature of inorganic foulants on the membrane or pore surface. For illustration, Figure 3.4a, b displays the EDX spectra of an alumina membrane layer before and after

Figure 3.3. Scanning electron micrograph of an alumina membrane surface (Hsieh, Bhave and Fleming 1988).

fouling by some compounds containing Fe, Si, Mg, Ca and Ba. It may be noted that gold appears in both spectra as a result of the use of gold as the coating material for sample preparation. Aluminum in the spectra is indicative of the membrane layer material, alumina.

Transmission electron microscopy (TEM) has higher resolution power than SEM and can be used to characterize some special membranes or their precursors. In the TEM, the electrons that form the image must pass through the specimen. This limits the thickness of the specimen to be analyzed to a few hundred nanometers in general, or to a few microns with more sophisticated TEMs. To characterize membranes with ultrafine pores in the nanometer range, TEM may prove to be useful in examining membrane precursor crystallites or particles (Figure 3.5a) or thin, unsupported membrane film (Figure 3.5b). Due to their mode of operation, TEM cannot reveal the morphological features of a relatively thick, composite membrane. As with SEM, elemental analysis can be performed down to the Angstrom range by TEM equipped with EDX.

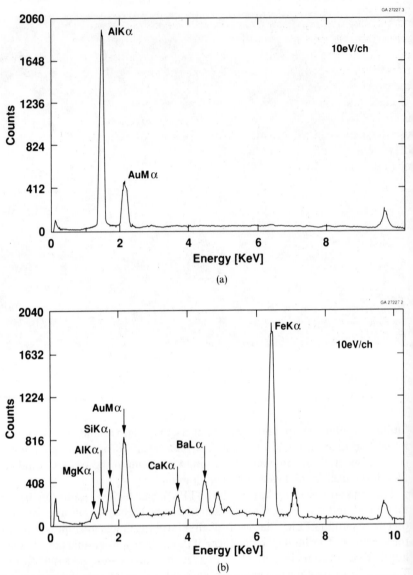

Figure 3.4. Energy-dispersive X-ray analysis spectra of an alumina membrane layer (a) before fouling and (b) after fouling.

3.3.2. Thickness

The thickness of the separative membrane layer for asymmetric membrane structures represents a trade-off between the physical integrity requirement, on the one hand, and the high flux requirement, on the other. Current commercial products show a membrane thickness of as thin as approximately 5 microns but generally in the 10 to 20 micron range. The bulk support and any intermediate support layers vary in thickness, but the bulk support needs to be thick enough (about 1–2 mm) to provide high mechanical strength. The intermediate layers range from 10 to 50 microns in thickness.

The quality of the support layer(s) beneath the separative membrane layer is critical to the quality of the membrane itself. A smooth membrane layer is usually associated with a smooth support layer beneath it and the interface between the membrane and the support is consequently sharp showing little or no blockage of pores by particles.

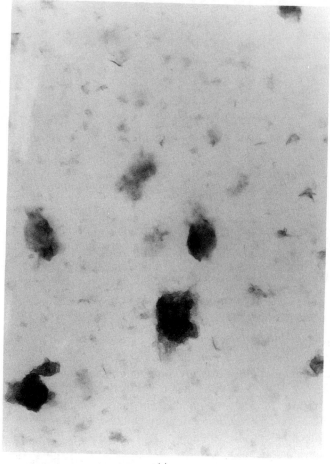

(a)

Figure 3.5. Transmission electron micrographs of (a) membrane precursor crystallites and (b) a thin, unsupported membrane film.

(b)

Figure 3.5 (*continued*)

Due to their preparation methods, inorganic membranes are in general structurally stable. They do not suffer from any appreciable dimensional instability problems due to compaction and swelling which commonly occur among many organic polymeric membranes.

3.3.3. Pore Size

Pore size plays a key role in determining permeability and permselectivity (or retention property) of a membrane. The structural stability of porous inorganic membranes under high pressures makes them amenable to conventional pore size analysis such as mercury porosimetry and nitrogen adsorption/desorption. In contrast, organic polymeric membranes often suffer from high-pressure pore compaction or collapse of the porous support structure which is typically "spongy".

The commercial mercury porosimeters can usually provide pore diameter distribution data in the range of 3.5 nm to 7.5 microns. It is a useful and commonly used method for characterizing porous particles or bodies. Figure

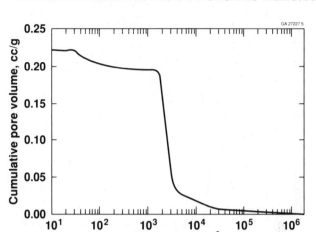

Figure 3.6. Pore size distribution by mercury porosimetry of a two-layered zirconia membrane composite.

3.6 describes the pore size distribution by mercury porosimetry of a two-layered zirconia composite membrane: a separative film with a pore diameter of approximately 4 nm and a single layer of support with a nominal pore diameter of 0.2 micron. Typical mercury porosimetry data come in two modes: intrusion and extrusion. The intrusion data are commonly used for two major reasons: (1) The intrusion step precedes the extrusion step in the mercury porosimetry analysis and (2) The complete extrusion of mercury out of the pores during the depressurization step of the analysis may take a very long time.

For membranes with pore diameters smaller than 3.5 nm, the nitrogen adsorption/desorption method based on the widely used BET theory can be employed. This measurement technique, however, is good only for pore diameters ranging from 1.5 nm to 100 nm (= 0.1 micron). Typical data from this method are split into two portions: adsorption and desorption. The nitrogen desorption curve is usually used to describe the pore size distribution and corresponds better to the mercury intrusion curve. Given in Figure 3.7 is the pore size distribution of an unsupported experimental membrane film. For composite membranes with the separative film and/or support having pores larger than 100 nm, the nitrogen adsorption/desorption method is not suitable. These membranes can be analyzed with mercury porosimeters as discussed above.

The task of analyzing thin asymmetric or graded-structure membranes with a wide range of pore sizes is quite challenging even when powerful tools

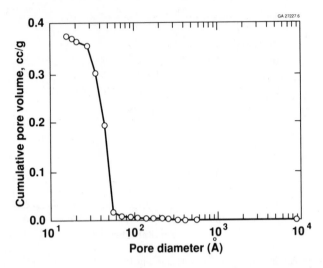

Figure 3.7. Pore size distribution by nitrogen desorption of an unsupported experimental alumina membrane film.

such as mercury porosimetry are used. The difficulty lies in the very small percentage of pore volume contributed by the thin membrane film relative to that of the support layer(s). A possible but cumbersome solution to the problem is to "shave" most of the bulk support layer to increase the pore volume percentage of the thin film and other thin layers of support. Knowing the amount of bulk support layer removed and the mercury porosimetry data of the "shaved" membrane sample, it is possible to combine the two pieces of information to arrive at the pore size distribution of a multiple-layered composite membrane. This is illustrated in Figure 3.8.

Another possible solution to the problem of analyzing multiple-layered membrane composites is a newly developed method using NMR spin-lattice relaxation measurements (Glaves 1989). In this method, which allows a wide range of pore sizes to be studied (from less than 1 nm to greater than 10 microns), the moisture content of the composite membrane is controlled so that the fine pores in the membrane film of a two-layered composite are saturated with water, but only a small quantity of adsorbed water is present in the large pores of the support. It has been found that the spin-lattice relaxation decay time of a fluid (such as water) in a pore is shorter than that for the same fluid in the bulk. From the relaxation data the pore volume distribution can be calculated. Thus, the NMR spin-lattice relaxation data on a properly prepared membrane composite sample can be used to derive the pore size distribution that conventional pore structure analysis techniques

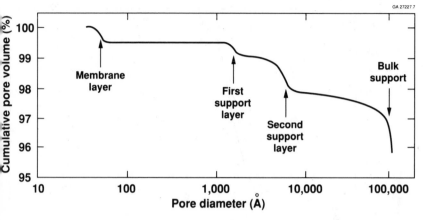

Figure 3.8. Pore size distribution of a four-layered alumina membrane (Hsieh, Bhave and Fleming 1988).

cannot easily provide. The size of the membrane sample to be analyzed by this experimental method is only limited by the homogeneity of the magnetic field. Currently, the practical limit is in the order of 10 cm, which is still much larger than the limits imposed by conventional pore size analyses such as mercury porosimetry or nitrogen adsorption/desorption.

Finally, another nondestructive method for ascertaining information on membrane pore size distributions has been developed and well tested at Oak Ridge Gaseous Diffusion Plant (Fain 1990). Different from all the previously mentioned techniques, this flow-weighted pore size distribution test method is based on gas transport rather than volume. It is not sensitive to the amount of gas adsorbed and particularly suitable for gas separation applications. The method consists in measuring the flow of a mixture of an inert gas (such as nitrogen) and a condensable gas (such as carbon tetrachloride) through membrane pores of various sizes. First, the gas mixture is pressurized to effect capillary condensation which completely plugs the pores. As the pressure is decreased incrementally, the flow is measured in the large pores first, followed by small pores. The pressure is decreased until there is no longer an increase in gas flow rate. At each pressure, the flow is measured. The change in flow rate between pressures is then related to the pore size through the Kelvin equation for capillary condensation.

The test is recommended to be operated with a small pressure difference (less than 3 cm Hg) across the membrane and a low mole fraction (0.05–0.1) of the condensable gas. The time to complete an analysis is dictated by the temperature and pressure equilibrium times which are typically a few hours. The test can accommodate membrane samples of various shapes and sizes

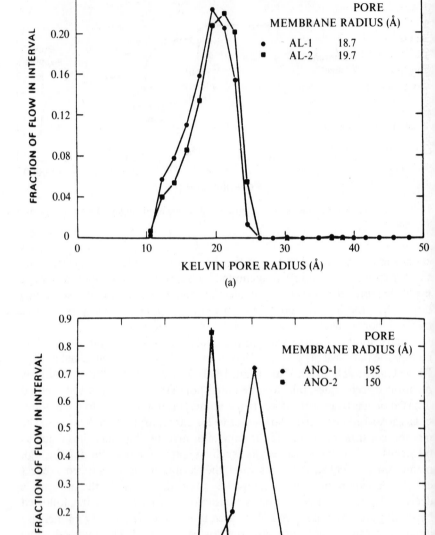

Figure 3.9. Dynamic pore size distributions of (a) two sol–gel alumina membranes and (b) two anodized alumina membranes (Fain 1990).

and is generally very reproducible (Fain 1990). For asymmetric or composite membranes, the generated data represent the distribution of smallest pores in the flow paths even though they may be only a small fraction of the volume of the membrane. Some test results for alumina membranes made by the sol–gel method and the anodic oxidation method are given in Figure 3.9a, b. The method is capable of measuring pore diameters up to approximately 1 micron and down to the 1.5–2.0 nm range where the Kelvin equation is valid.

3.3.4. Permeabilities and Retention Properties

The pore size of a membrane is manifested in the permeabilities and separation (retention) characteristics of the membrane. For gases, the pore size of the membrane also affects the prevailing transport mechanisms through the pores (see Chapter 6).

Typically, the liquid permeabilities are obtained with water being the permeate and expressed in terms of L/h-m²-bar. The gas permeabilities are often expressed in terms of air or nitrogen permeabilities. The retention characteristics can be generally and generically obtained by using some model molecules. The most commonly used model molecules are polyethylene glycol (PEG) polymers which are linear and flexible in nature, and

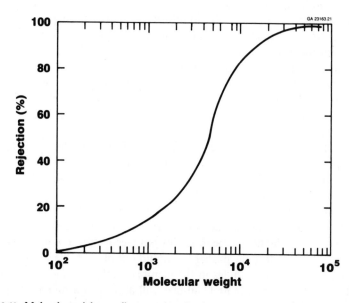

Figure 3.10. Molecular weight cutoff curves of an alumina membrane using dextrans as the model molecule (Leenaars and Burggraaf 1985b).

dextrans which are slightly branched. These model molecules are widely used primarily due to their relatively low cost. They are, however, far from being perfectly spherical and, as such, the molecular weight cutoff curves using these model molecules are not ideally sharp. Figure 3.10 shows the molecular weight cutoff curves of two alumina membranes using PEG as the model molecule (Leenaars and Burggraaf 1985b). The nonideality of the molecular weight cutoff is a result of the combination of linear and flexible molecular shape. Other materials such as latex particles can be used as the model system. These effects are more thoroughly discussed in Chapter 4.

3.3.5. Maximum Pore Size and Structural Defects

A commonly used simple method for determining if there are any cracks or pinholes in microporous membranes is the so-called bubble point test. It has been used by many organic membrane manufacturers and users alike and is also being adopted by some inorganic membrane manufacturers. The method utilizes the Washburn equation

$$d = 4S\cos\theta/\Delta P \qquad (3.1)$$

where d is the pore diameter, S the surface tension of the liquid used, θ the contact angle between the membrane material and the liquid, and ΔP the applied pressure difference. It essentially states that a given pressure difference is required to displace a liquid (having a surface tension S and a contact angle θ with the pore surface) from a pore having a diameter "d" with a gas such as air or nitrogen. A frequently used liquid medium is water. However, alcohols or hydrocarbons have also been used due to their lower surface tension (e.g. surface tension of some fluorocarbon liquids is only 16 dyne/cm compared to 72 dyne/cm for water) which results in a lower pressure difference required to displace the liquid.

This method has been approved as an ASTM procedure (F316). The typical apparatus used is schematically illustrated in Figure 3.11. An instrument based on this method has been introduced by Coulter Electronics recently (Venkataraman et al. 1988).

The bubble point test is most often used for detecting the largest pore size of the membrane by finding the ΔP [thus the pore diameter according to Equation (3.1)] at the first appearance of bubbles from the liquid-saturated membrane when the test gas pushes the liquid out of the largest size pores. If there are any cracks or pinholes in the membrane, the method will see them as the largest pores and the first bubbles will appear at a much lower than usual pressure. In fact, the bubble point test was originally devised as a simple quality control tool for checking the physical integrity of membranes.

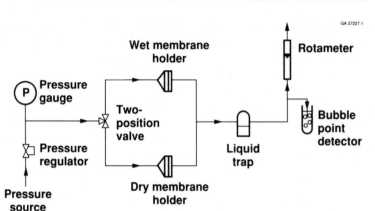

Figure 3.11. Schematic of a bubble point test apparatus.

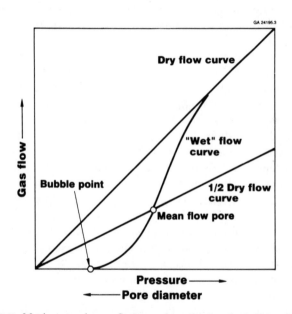

Figure 3.12. Maximum and mean flow pore determinations by bubble point test.

This method in principle can also be used to estimate the "average" pore size of the membrane when there are no defects in the membrane. First, gas flows through the membrane in the dry state. Typically, the gas flow rate is a linear function of the applied pressure difference. Then the membrane is saturated with the test liquid and the gas is forced to flow through the wet membrane. When the first bubbles appear, the corresponding pressure is called the bubble point pressure (Figure 3.12). If the test continues beyond the bubble point, as indicated in Figure 3.12, the "wet" curve will approach the "dry" curve at a relatively high pressure (usually several bar). The pressure at which the "half-dry" curve (those points with half the dry flow rate) intersects with the "wet" curve can be used to calculate the "average" pore diameter according to Equation (3.1).

3.3.6. Characteristics of Pore Network

The size of the crystallites forming the network of membrane pores and the porous nature of the network affect the permeation and separation properties of porous inorganic membranes. These features will be discussed below.

From the mercury porosimetry data, porosity can be calculated. A higher porosity means a more open pore structure, thus generally providing a higher permeability of the membrane. Porous inorganic membranes typically show a porosity of 20 to 60% in the separative layer. The porous support layers may have higher porosities.

The size of crystallites that make up inorganic membranes with narrowly distributed pores in the nanometer range can be estimated by using transmission electron microscopy (TEM). However, TEM involves tedious sample preparation procedures to reveal information on crystallites in a membrane layer. X-ray diffraction line-broadening measurements, when implemented in a computerized X-ray diffraction system, can be used to estimate the crystallite size. The technique is based on the Scherrer equation (Klug and Alexander 1974). When there are no significant amounts of lattice strains or other imperfections in a solid sample, the breadth of the pure X-ray diffraction profile can be ascribed solely to small crystallite size. Appropriate theoretical relationships may be used to estimate the mean crystallite dimension and, in some cases, limited information about the size distribution may be obtained as well. The technique has been found to provide crystallite sizes fairly close to those determined from TEM. For example, the crystallite diameters of several experimental unsupported transition-alumina membranes were estimated to be in the range 2.8–4.4 nm by the X-ray diffraction line-broadening technique. For comparison, TEM determined the crystallite diameter to be 4–5 nm.

The internal surface area of a porous inorganic membrane is often significantly affected by the heat treatment temperature. Leenaars, Keizer and Burggraaf (1984) have shown that, even if the crystallite size of the membrane precursor particles remains essentially the same (from the X-ray line-broadening measurements), the surface area of a transition-phase alumina membrane decreases with increasing calcination temperature. Consequently, the mean pore diameter of the resultant membrane increases with the calcination temperature.

Depending on the preparation method used, the path of the membrane pores can be nearly straight or tortuous. The primary particles and their aggregates which are used to form porous membranes can greatly vary in morphology as a result of synthesis and calcining conditions. The resultant pore networks have varying degrees of randomness. A commonly used empirical approach to describing the degree of randomness of the pore structure in a membrane is the factor "tortuosity" which is determined from the permeability, internal surface area, pore volume fraction, pore shape and thickness of the membrane and the permeate viscosity. The Kozeny–Carmen equation (See Chapter 4), which has been widely used to describe laminar flow through porous media (such as membranes), can then be used to calculate the tortuosity of the membrane pores (Leenaars and Burggraaf 1985a). It has been demonstrated that the tortuosity value determined this way agrees well with that obtained by the measurement of effective ionic conductivity (Shimizu et al. 1988). However, due to variabilities of the parameters involved, tortuosity has only been referred to in the scientific literature and has not been used by either manufacturers or users to describe the membrane pore structures or associated transport properties.

3.4. MATERIALS PROPERTIES

3.4.1. Chemical Resistance

In general, inorganic membranes are inherently more stable to various chemicals over a wider range of pH values than organic membranes. The stability of these inorganic materials, particularly ceramic materials, is the result of the compactness of the crystal structure, the chemical bonding and the high field strengths associated with the relatively small and highly-charged cations in ceramics. From a thermodynamic viewpoint, the stable ceramics will be those having a large free energy and a large total energy of formation. Therefore, yttria and thoria are among the most stable in this respect. Alumina, beryllia, titania and zirconia are also recognized for their chemical stability.

Nonetheless, for situations where application environments or membrane regeneration (e.g. cleaning) procedures call for long contact between corrosive chemicals such as strong acids or bases and the inorganic membranes, the issue of potential chemical attack needs to be addressed. Also, it is generally known that nonoxide ceramic materials will not tolerate long-term exposure to oxidizing environments, particularly at high temperatures.

Various ceramic membranes, for example, possess differing degrees of acid/base resistance, depending on the pH value, particular phase of the membrane material, porosity, contact time and temperature. However, no quantitative data are available on the kinetics of chemical dissolution of ceramic membranes as a guide for chemical corrosion considerations.

Porous glass membranes show excellent resistance to acid attack, but exhibit some degree of dissolution in the presence of strong caustic solutions, particularly at elevated temperatures. They are also known to undergo some structural changes upon long exposure to water due to partial dissolution of silica. Different phases of a given metal oxide such as alumina can also exhibit markedly different chemical reactivities in wet systems. For example, as a general rule, as the density of an alumina (and hydrous alumina) phase increases, its chemical resistance increases. The hexagonal α-alumina crystal shows very good resistance to chemical attack including strong acids or bases. In contrast, the cubic crystalline transition alumina with a pore diameter of a few nanometers does not possess good resistance to highly acidic or basic media.

There are limited data on the chemical resistance of various oxide materials in the literature (Samsonov 1982, Ryshkewitch and Richerson 1985). Furthermore, many of the studies are concerned with solid, nonporous materials. Nevertheless, these may provide an indication of the general trends. Until a definitive and quantitative database of chemical stability of various inorganic membranes becomes available, some simple dissolution-type test methods using membrane samples may be employed on a comparative basis to estimate the extent of attack by a chemical under the application conditions. An example of such a simple test is given below.

Various membrane materials are to be compared for corrosion resistance in hydrochloric acid. Membrane samples are ultrasonically cleaned with Freon for 5 minutes and dried at 200°C for 2 hours followed by similar steps of ultrasonic cleaning with demineralized water and drying. The conditioned membrane samples are then immersed in 35% HCl solution, making sure that no air bubbles are trapped in pores. The acid exposure at the test temperature (e.g. 25°C) continues for a given period (e.g. one week). The tested samples are ultrasonically washed with demineralized water for 5 minutes and dried at 200°C for 2 hours. The weights of the cleaned membrane samples before and after the acid exposure are compared to assess the relative corrosion resistance of various membrane materials.

Some test data are shown in Table 3.2 where alumina is compared to PTFE and stainless steel 316. Since corrosion is predominantly initiated at the fluid–solid interface, it seems plausible to compare the corrosion results on a per unit surface area basis. In the above example, from the data of weight change per unit area per day or % weight change per unit area per day,

Table 3.2. Test Data for Comparing HCl Corrosion of Various Membrane Materials

Material	Surface Area (cm^2/g)	Wt. Loss % per unit cm^2 per day
α-Alumina	0.93	2.1×10^{-5}
PTFE	3.21	6.4×10^{-4}
Stainless steel. 316(A)	3.09	4.1×10^{-1}
Stainless steel 316(B)	0.85	4.3×10^{-2}

GA 27227 4

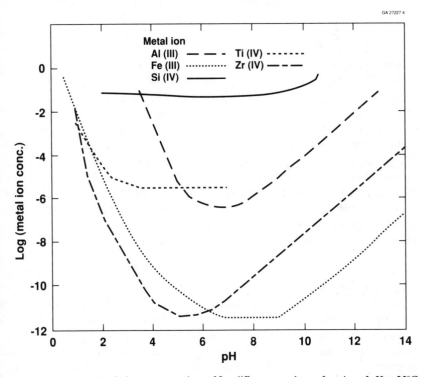

Figure 3.13. Saturated solution concentrations of five different metals as a function of pH at 25°C.

alumina membrane is seen to exhibit better HCl resistance than PTFE and stainless steel 316.

Some trends on the long-term behavior of the membrane to corrosion by acids and bases may be obtained through thermodynamic considerations. Shown in Figure 3.13 is a comparison of saturated solution concentrations of five different metals as a function of pH at 25°C. From the equilibrium data of Figure 3.13, the tendency for various metal oxides to be attacked by caustic or acid solutions can be revealed, at least on a relative scale. It is seen that silica, on a relative basis, will be dissolved more than other oxides over a long period and under a wide range of pH. In the absence of kinetic data on chemical corrosion of actual membrane samples, the figure sheds some light on what to expect in terms of relative pH resistance. It should be emphasized that Figure 3.13 cannot be used for performance prediction but only serves for purposes of comparison.

It is noted that finely divided oxides as in the case of membrane structure can be rapidly dissolved in HF, hot concentrated sulfuric acid, ammonium fluoride, and concentrated hydrochloric acid, especially when under pressure.

3.4.2. Surface Properties

It is generally recognized that the surface charges of a membrane, which are largely dependent on the membrane material and operating conditions, can significantly influence the separation performance of the membrane in terms of both flux and permselectivity. For example, the permeate flux decline observed almost invariably depends strongly on the adsorption of solute components on the membrane surface. This decline in flux, often called fouling, is usually promoted by ions in the process streams. This aspect is further discussed in Chapters 4 and 5. The electrostatic component of the interaction forces between the molecules in solution and the surface layer of the membrane depends to a great extent on the charge density of the membrane surface and its pores (Nystrom, Lindstrom and Matthiasson 1989). The surface charge density is characterized by the zeta potential of the membrane.

An indirect way of studying the zeta potential of the membrane is by using the membrane precursor particles or grinding the membrane (or film) material into particles which are then amenable to zeta potential measurement. For example, adsorption of surfactants onto glass has been studied through the determination of zeta potential and surface free energy of ground membrane material (Busscher et al. 1987). Similarly, surface modifications to alumina membranes with silanes and a sulfone (Shimizu et al. 1987) and with trimethyl chlorosilane TMS (Shimizu et al. 1989) have been shown to alter the zeta potential of the membranes by characterizing the membrane precursor particles as a substitute for the membranes. Further, it has also been shown that zeta potential under normal membrane operating conditions varies little even over a period of 2–3 years (Shimizu et al. 1989). The effect of surface properties on the filtration performance is discussed in Chapter 4. The pore size distribution variability has negligible effects on surface charges. This type of indirect measurement approach can be complicated by the sample preparation (i.e. grinding) step itself as excess surface charges due to grinding are known to exist in the study of comminution.

Other methods have been investigated to characterize surface charges of inorganic membranes. Schwarz et al. (1986) have measured the so-called concentration potential which has been shown to affect the filtrate flux through a pressure-driven polysulfone membrane. In ideal cases where the membrane pores are straight, an electrokinetic phenomenon called streaming potential (as a result of liquid flow through capillaries or pores under a pressure difference) can be used to calculate the zeta potential. Streaming potential can be a physically significant parameter indicative of the interaction between membranes and liquid. Its reliable determination, however, is not easy (Adamson 1982). Nystrom, Lindstrom and Matthiasson (1989) have devised a sophisticated apparatus for measuring streaming potentials. They used the experimentally determined streaming potentials to calculate the apparent zeta potentials of membranes by applying the Helmholtz–Smoluchowski equation. The method is found to provide consistent results and gives valuable information for the prediction of adsorption, separation properties and fouling, especially for solutions containing ionic macromolecules. Even though the method was applied to organic membranes, it should also be applicable to inorganic membranes of various pore sizes.

Interactions between the membrane and the liquid can be caused by various mechanisms such as ionic and van der Waals interactions and hydrogen bonding. Surface modifications have become an effective and sometimes economic means of changing the interactions and, hence, altering the separation properties. In recent years there has been an increasing amount of research on surface-modified inorganic membranes (Schnabel and Vaulont 1978, Messing 1979, Shimizu et al. 1987, Keizer et al. 1988, Yazawa et al. 1988).

3.4.3. Mechanical Properties

Despite the fact that inorganic membranes are, in general, more stable mechanically than organic membranes, available mechanical properties data for commercial inorganic membranes are sketchy and these are not yet standardized for comparing various membranes. It appears that the methods used for obtaining various mechanical strength data are based on those for solid (nonporous) bodies and most of them are listed as ASTM procedures.

Tabulated in Table 3.3 are the various mechanical properties taken from product brochures of commercial inorganic membranes. The table is not intended for comparing different membranes as the reported data may not be obtained under similar test conditions. However, it is expected to give at least

Table 3.3. Mechanical Property Data of Various Inorganic Membrane Elements
from Product Brochures

Membrane Material	Burst Strength (bar)	Bending Strength (bar)	Flexural Strength (bar)	Compressive Strength (bar)
Al_2O_3 (Alcoa)	> 100			
Al_2O_3 (NGK)	> 60	50–450		
ZrO_2 (SFEC)	60			
Glass (Asahi)	40		400–700	2,050
Ag (Osmonics)				15,000

an order-of-magnitude estimate in the absence of standard procedures. It is also indicative of the lack of systematic approach to the issue of mechanical properties. As inorganic membranes are more widely used, the applications data base should reveal the important parameters to be closely studied.

3.5. MEMBRANE ELEMENT AND MODULE CONFIGURATIONS

The structural elements of commercial inorganic membranes exist in three major geometries: disk, tube or tube bundle, and multichannel or honeycomb monolith. The disks are primarily used in laboratories where small-scale separation or purification needs arise and the membrane filtration is often performed in the flow-through mode. The majority of industrial applications require large filtration areas (20 to over 200 m²) and, therefore, the tube/tube bundle and the multichannel monolithic forms, particularly the latter, predominate. They are almost exclusively operated in the cross-flow mode.

Figure 3.14 shows membrane elements in both tubular and monolithic geometries. The single-tube configuration is often used in laboratory feasibility studies. The flow path of permeate through a porous multichannel monolithic membrane element is depicted in Figure 3.15. The monolithic form offers the advantages of easier installation and maintenance and more filtration area per unit volume of the membrane element (which affects the process economics) and is also more robust mechanically than the tube bundle configuration. Multichannels with small channel diameters (3–6 mm) also consume less energy for pumping the feed streams through the filtration loop. In some large-scale applications the number of tubes in a module may exceed a thousand, as in the case of some zirconia membrane systems (Gaston County Filtration Systems 1980).

A membrane module may contain many membrane elements as shown in Figure 3.16, where 19 monolithic membrane elements each with 19 channels

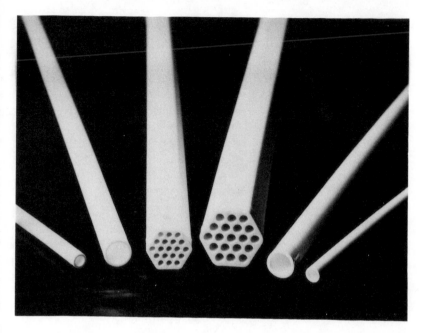

Figure 3.14. Photograph of alumina membrane elements (Alcoa 1987).

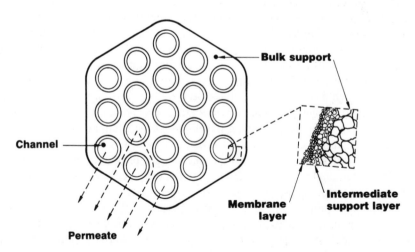

Figure 3.15. Schematic of permeate flow path through a porous multichannel monolithic membrane element (Hsieh 1988, Hsieh, Bhave and Fleming 1988).

Figure 3.16. Photograph of membrane modules (Alcoa 1987).

are packed in a module. Thus, a significant amount of filtration area per unit volume of the module can be attained. Finally, a complete system of inorganic membranes may consist of many aforementioned membrane modules. Figure 3.17 shows an industrial-scale ceramic membrane system.

Process economics considerations have forced manufacturers to continuously improve the available filtration area per unit membrane element or module volume. For organic membranes, two unique membrane configurations with high packing densities have been well developed: hollow-fiber and spiral-wound. While there are no commercial spiral-wound inorganic membranes currently, there have been developments in preparing experimental hollow-fiber ceramic membranes (Baker et al. 1978, Beaver 1986). Commercially available inorganic membranes show a packing density of about 30 m^2/m^3 for single plates, 30–250 m^2/m^3 for tubes and 130–400 m^2/m^3 for multichannel monoliths. For hollow-fiber membranes, the packing density can be greater than 1000 m^2/m^3. Recently Ceramem Corporation (Waltham, MA) have commercialized porous ceramic microfiltration and ultrafiltration membranes deposited on honeycomb monolith coerdierite supports (Goldsmith 1988, 1990). Packing density as large as 800 m^2/m^3 can be achieved on full-size modules.

Figure 3.17. A typical fruit juice plant using Membralox® ceramic membrane filters (approx. 100 m² membrane area). (Courtesy: MEMBRAFLOW Filter systems, Germany).

3.6. END-SEAL AND MODULE PACKING MATERIALS

The ends of membrane tubes or monoliths need to be impermeable to all components to prevent the permeate from remixing with the retentate(s), or the retentate(s) from leaking to the permeate side through any imperfections in the separative layer or in large support pores at the ends of the membrane elements. The long-established ceramic and glass technologies can provide the needed solutions to this problem. Coating the ends of the membrane element with various enamels or with very fine ceramic particles are just some examples. An important issue in selecting an effective end-seal material is matching the thermal expansion coefficients between the membrane element and the sealing material.

The materials used for packing membrane elements in a module are critical in relation to the application conditions. For liquid phase processing, polymeric materials such as silicone rubber, epoxy, polyester, Viton, Teflon and Buna N have been used for connecting some ceramic membranes to the housing or header plates of the module which are made of stainless steel or plastic (such as CPVC). For high-temperature applications (particularly gas phase processing), several approaches have been attempted. High-temperature polymers such as PTFE and aramid fiber can be used up to about 250°C. Other polymers, polyimides, polyetherimides and polyamideimides are known to withstand temperatures greater than 230°C for a reasonable period of time. Graphite/carbon filament and flexible graphite packings may be used up to about 300–400°C under oxidizing conditions and up to 1000°C under reducing or inert environments. Ceramic to metal (cermet) connections are also a possibility. Other approaches include localized cooling near the packing areas and compressible Knudsen flow-type seals (Nourbakhsh et al. 1989). Attention should be given to such factors as their thermal stability, chemical compatibility and difference in thermal expansion coefficient with the membrane elements (to avoid cracking problems).

REFERENCES

Adamson, A. W. 1982. *Physical Chemistry of Surfaces*. 4th ed. John Wiley & Sons, New York.
Alcoa. 1987. Membralox® ceramic multichannel membrane modules. Product brochure.
Baker, R. A., G. D. Forsythe, K. K. Likhyani, R. E. Roberts and D. C. Robertson. 1978. Separation device of rigid porous inorganic hollow filament and use thereof. U. S. Patent 4,105, 548.
Ballou, E. V., M. I. Leban and T. Wydeven. 1973. Solute rejection by porous glass membranes. III. Reduced silica dissolution and prolonged hyperfiltration service with feed additive. *J. Appl. Chem. Biotechnol.* 23: 119–30.
Bansal, I. K. 1976. Concentration of oily and latex waste waters using ultrafiltration inorganic membranes. *Ind. Water Engr.* 13 (5): 6–11.

Beaver, R. P. 1986. Porous hollow silica-rich fibers and method of producing same. European Patent Appl. 186,129.

Busscher, H. J., H. M. Uyen, G. A. M. Kip, and J. Arends. 1987. Adsorption of aminefluorides onto glass and the determination of surface free energy, zeta potential and adsorbed layer thickness. *Colloids and Surfaces* 22: 161–69.

Fain, D. E. 1990. A dynamic flow-weighted pore size distribution. *Proc. 1st Intl. Conf. Inorganic Membranes*, 1–5 July 1989, 199–205, Montpellier.

Gaston County Filtration Systems. 1980. Gaston County ultrafiltration systems. Product brochure.

Gerster, D, and R. Veyre. 1985. Mineral ultrafiltration membranes in industry. In *Reverse Osmosis and Ultrafiltration*, eds. S. Sourirajan and T. Matsuura, pp. 225–30. Washington, D.C.: Am. Chem. Soc.

Gillot, J. and D. Garcera. 1984. New ceramic filter media for cross-flow microfiltration and ultrafiltration. Paper presented at Filtra 84 Conference, 2–4 October 1984, Paris.

Gillot, J., R. Soria, and D. Garcera. 1990. Recent developments in the Membralox[R] ceramic membranes. *Proc. 1st. Intl. Conf. Inorganic Membranes* 3–6 July pp. 379–81, Montpellier.

Glaves, C. L., P. J. Davis, K. A. Moore, D. M. Smith and H. P. Hsieh. 1989. Pore structure characterization of composite membranes. *J. Colloid and Interface Sci.* 133(2): 377–89.

Goldsmith, R. L. 1988. Cross-flow filtration device with filtrate flow conduits and method of forming same. U. S. Patent 4,781,831.

Goldsmith, R. L. 1990. Low-cost ceramic membrane modules. Paper read at 1990 Eighth Annual Membrane Technology Planning Conference, October 16, Newton, MA.

Hsieh, H. P. 1988. Inorganic membranes. In AIChE Symp. Ser. *New Membrane Materials and Processes of Separation*, eds. K. K. Sirkar and D. R. Lloyd, pp. 1–18. New York: Am. Institute of Chem. Engr.

Hsieh, H. P., R. R. Bhave and H. L. Fleming. 1988. Microporous alumina membranes. *J. Membrane Sci.* 39: 221–41.

Kameyama, T., K. Fukuda, M. Fujishige, H. Yokokawa and M. Dokiya and Y. Kotera. 1981. Production of hydrogen from hydrogen sulfide by means of selective diffusion membranes: *Adv. Hydrogen Energy Prog.* 2: 569–79.

Keizer, K., R. J. R. Uhlhorn, R. J. van Vuren and A. J. Burggraaf. 1988. Gas separation mechanisms in microporous modified gamma-Al_2O_3 membranes. *J. Membrane Sci.* 39: 285–300.

Kluġ, H. P., and L. E. Alexander. 1974. *X-ray Diffraction Procedures for Polycrystalline and Amorphous Materials.* 2nd ed. John Wiley & Sons, New York.

Leenaars, A. F. M., K. Keizer and A. J. Burggraaf. 1984. The preparation and characterization of alumina membranes with ultrafine pores. Part 1. Microstructural investigations on non-supported membranes. *J. Mat. Sci.* 19: 1077–88.

Leenaars, A. F. M. and A. J. Burggraaf. 1985a. The preparation and characterization of alumina membranes with ultrafine pores: 2. The formation of supported membranes. *J. Coll Interface Sci.* 105(1): 27–40.

Leenaars, A. F. M. and A. J. Burggraaf. 1985b. The preparation and characterization of alumina membranes with ultrafine pores. Part 4. Ultrafiltration and hyperfiltration experiments. *J. Membrane Sci.* 24: 261–70.

Messing, R. A. 1979. Hydrophobic inorganic membrane for gas transport. U.K. Patent 2,014,868A.

Nourbakhsh, N., A. Champagnie, T. T. Tsotsis and I. A. Webster. 1989. Transport and reaction studies using ceramic membranes. In A.I.Ch.E. Symp. Ser. *Membrane Reactor Technology*, eds. R. Govind and N. Itoh, pp. 75–84. New York: Am. Institute of Chem. Engr.

Nystrom, M., M. Lindstrom and E. Matthiasson. 1989. Streaming potential as a tool in the characterization of ultrafiltration membranes. *Colloids & Surfaces* 36: 297–312.

Ryshkewitch, E. and D. W. Richerson. 1985. *Oxide Ceramics—Physical Chemistry and Technology.* 2nd ed. General Ceramics, Inc., New Jersey.

Samsonov, G. V. 1982. *The Oxide Handbook.* 2nd ed. IFI/Plenum Data Co., New York.

Schnabel, R. and W. Vaulont. 1978. High-pressure techniques with porous glass membranes. *Desalination* 24 (1–3): 249–72.

Schwarz, H., V. Kudela, J. Lukas, J. Vacik and V. Grobe. 1986. Effect of the membrane potential on the performance of ultrafiltration membranes. *Collection Czechoslovak Chem. Commun.* 51: 539–44.

Shimizu, Y., T. Yazawa, H. Yanagisawa and K. Eguchi. 1987. Surface modification of alumina membranes for membrane bioreactor. *Yogyo Kyokaishi* 95: 1067–72.

Shimizu, Y., K. Matsushita, I. Miura, T. Yazawa and K. Eguchi. 1988. Characterization of pore geometry of alumina membrane. *Nippon Seramikkusu Kyokai Gakujutsu Ronbunshi* 96(5): 556–60.

Shimizu, Y., K. Yokosawa, K. Matsushita, I. Miura, T. Yazawa, H. Yanagisawa and K. Eguchi. 1989. Zeta potential of alumina membrane. *Nippon Seramikkusu Kyokai Gakujutsu Ronbunshi* 97(4): 498–501.

Venkataraman, K., W. T. Choate, E. R. Torre, R. D. Husung and H. R. Batchu. 1988. Characterization studies of ceramic membranes: A novel technique using a Coulter porometer. *J. Membrane Sci.* 39(3): 259–71.

Yazawa, T., H. Nakamichi, H. Tanaka and K. Eguchi. 1988. Permeation of liquid through a porous glass membrane with surface modification. *Nippon Seramikkusu Kyokai Gakujutsu Ronbunshi* 96(1): 18–23.

4. Permeation and Separation Characteristics of Inorganic Membranes in Liquid Phase Applications

R. R. BHAVE

Alcoa Separations Technology, Inc., Warrendale, PA

4.1. INTRODUCTION

The principle of separation using microfiltration (MF) and ultrafiltration (UF) is based on the concept of size exclusion or sieving. The transport through porous inorganic membrane barriers occurs through the inter-granular spaces (and not through the granular particles themselves) within the top membrane layer, porous sublayers (in the case of asymmetric membrane structure) and the porous support structure (Hsieh 1988, Hsieh, Bhave and Fleming 1988). In 'contrast, the transport across polymeric membrane structures occurs through the continuous network of openings (i.e. the actual pores) from one face to the other. In practice, however, neither of these idealized pore structure is useful to completely predict or interpret the observed filtration performance. This is due to the occurrence of secondary phenomena such as interaction of solute or solvent with the membrane material and concentration polarization.

For a rational system design, it is therefore essential to have as much detailed knowledge of transport and separation mechanisms as possible to maximize the flux across an inorganic membrane barrier. The basic perme-ation and separation characteristics are reviewed here and a mathematical analysis is presented for the calculation of many important system variables, such as permeate flux, dissolved solute or particulate concentration, solute molecular diameter, boundary-layer thickness, rejection coefficient and mass transfer coefficient.

4.2. COMMON TERMINOLOGY AND DEFINITIONS

Certain terms often used in this chapter are defined below.

Cake layer A fouling layer resulting primarily from the physical deposition of solids on the membrane surface.

Concentration polarization This occurs when solutes accumulate on or near the membrane surface as a result of their exclusion from the permeate.

Concentration factor The ratio of final component concentration to its initial concentration.

Cross-flow filtration This is a pressure-driven separation process in which the feed stream flows parallel to the filter and generates two outgoing streams commonly identified as permeate and retentate.

Cross-flow velocity This is the average rate at which the process fluid flows parallel to the membrane surface.

Flux Permeation rate per unit filtering surface.

Gel layer A commonly used term for an extreme case of concentration polarization. It refers to the consolidated layer of concentrated solute on the membrane surface. It also describes a type of fouling layer commonly caused by gel-forming substances.

Internal pore fouling The deposition of material inside the porous structure, possibly leading to a reduction in pore volume.

Membrane fouling It is the manifestation of the accumulation (usually excessive) of particulates, solutes, colloids, etc., on the membrane surface.

Permeate The filtrate crossing the filtering surface.

Permeability Flux per unit transmembrane pressure.

Pore blockage This term relates to the reduction in the area available for permeation as a result of the actual physical external blockage of the pore by the foulant.

Retentate The concentrate which does not cross the filtering surface.

Transmembrane pressure (ΔP_T) It is the average pressure difference across the membrane structure from feed (or reject) side to permeate side.

Volumetric concentration factor The ratio of initial volume to retentate volume.

4.3. PHENOMENA INFLUENCING FLUX AND SEPARATION DURING THE FILTRATION PROCESS

During the course of a typical permeation and separation process using membranes a variety of phenomena such as concentration polarization, adsorption and internal pore fouling influence the permeate rate. In particulate filtration applications, pore fouling can be minimized by the proper selection of membrane pore size and operating conditions such as dynamic countercurrent backflushing. Nevertheless, in almost all practical situations these and other factors influence the flux and separation performance.

4.3.1. Concentration Polarization

Figure 4.1 shows schematically the phenomenon of concentration polarization. At a given temperature, feed rate and applied pressure difference, the permeate crossing the surface carries a certain amount of solute. This can be represented by the product of flow rate per unit area $(J = F/A)$ and solute concentration (C). As a result of solute transport across the membrane structure, the concentration of solute near the membrane surface can be higher than in the bulk. However, due to the higher solute concentration near the wall, there will be backdiffusion of solute from near the wall into the bulk liquid phase. This can be expressed as $D(dC/dy)$. Thus, at steady state

$$JC = D\frac{dC}{dy} \qquad (4.1)$$

Integrating,

$$\frac{C_w}{C_b} = \exp\left(J\frac{\delta}{D}\right) \qquad (4.2)$$

The mass transfer coefficient, k, is given by

$$k = \frac{D}{\delta} \qquad (4.3)$$

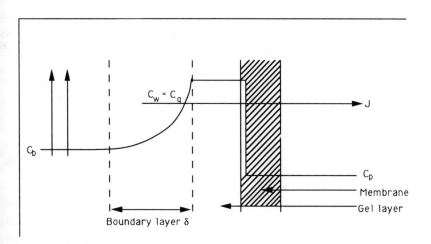

Figure 4.1. A schematic representation of the phenomenon of concentration polarization

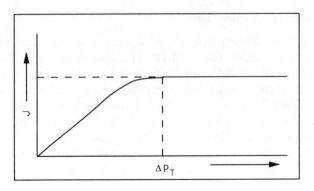

Figure 4.2. Typical dependence of permeate flux on transmembrane pressure.

The permeate flux, J, is obtained from

$$J = k \ln\left(\frac{C_w}{C_b}\right) \tag{4.4}$$

The value of permeate flux is influenced by a number of hydraulic parameters such as cross-flow velocity, wall roughness and solution viscosity (and hence temperature). When the wall concentration, C_w, reaches the concentration, C_g, at which gel formation occurs, the permeate rate is given by

$$J = k \ln\left(\frac{C_g}{C_b}\right) \tag{4.5}$$

The permeate flux under these conditions is strongly influenced by the backdiffusion of solute from the gel layer into the bulk feed. The permeate flux can be increased by increasing the value of k, by decreasing the gel layer thickness or less frequently, by increasing the diffusion coefficient through an increase in temperature. In practice, this is often accomplished by increasing the value of cross-flow velocity within permissible limits [e.g. constraints imposed by pressure drop (ΔP) considerations]. The above analysis indicates that under gel polarization conditions, transmembrane pressure can no longer influence the permeate flux due to the insensitivity of C_g (to pressure variations), which is essentially controlled by mass transfer parameters. This is illustrated in Figure 4.2.

4.3.2. Adsorption

Almost all macromolecules show an interaction with the surfaces in contact. This phenomenon, commonly described as fouling is due to adsorption; it is very complex and a fundamental study is often required to understand and predict its influence on flux reduction during filtration. To simplify the analysis, a monomolecular adsorption layer is assumed to be formed covering the membrane surface during contact with the feed solution (with or without permeation across the membrane surface area). A significant flux reduction (20–35%) can occur even with a monomolecular adsorption layer. The tendency to form adsorbed layer on the membrane surface may be a function of the nature of the membrane surface. For example, it is commonly observed that hydrophilic surfaces adsorb less strongly than hydrophobic surfaces especially when organics are involved.

It is known that the presence of adsorption layers modify the linear relationship between the flux and ΔP_T (Charpin et al. 1984). This is illustrated in Figure 4.3. In these situations, the difference, Δa, in the linear portions of the two F vs. ΔP_T relationships shown in Figure 4.3, may be attributed to the adsorption phenomenon occurring on the pore walls. In general, the smaller the pore size the larger will be the value of Δa, indicating that fouling due to adsorption is more likely in UF than in MF where the membrane pore sizes are larger (0.1–10 μm).

At high solute concentrations, however, a number of layers progressively build up on the membrane surface in contact. These secondary layers are responsible for further flux reduction which can be quite pronounced, and up to 90% (in some cases almost total flux loss) of the original flux can be lost.

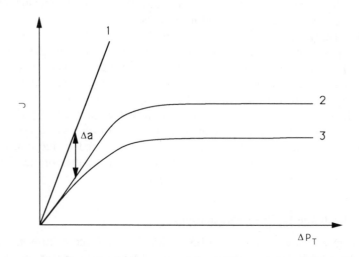

Figure 4.3. Typical flux vs. transmembrane pressure dependence for pure liquids and process liquids illustrating the presence of adsorption (Charpin et al. 1984). (1: water; 2,3: process feeds at two different flows, temperature or solute concentrations; Δa may be indicative of the influence of adsorption.)

The phenomenon of adsorption and fouling in the processing of milk using 0.2 μm alumina membranes was characterized with the help of scanning electron microscopy (SEM) and physicochemical analysis (Vetier, Bennasar and Tarodo de la Feunte 1988). In the processing of milk, soluble proteins and calcium and phosphorus salts accumulate on the membrane surface in the form of a fouling layer. This occurs due to the retention of serum milk proteins by the porous micellar deposit resulting in progressive fouling (which is generally a time-dependent phenomenon) of the membrane surface. It was found that the presence of Ca and P salts probably increased fouling due to the preferential adsorption of casein micelles on the alumina surface. This is shown in Figure 4.4. With the help of SEM and a physicochemical analysis, the rapid formation of the fouling layer was demonstrated.

The mechanism is postulated to begin with the adsorption of a thin film of casein and salts on which other micelles are then deposited. These may be connected by phosphocalcic bonds which form a porous layer that is largely responsible for the retention of serum proteins. This signifies the importance of surface adsorption phenomenon. It was also clearly shown that there was almost no penetration of casein micelles in the top layer, under typical cross-flow filtration conditions. The cross-flow velocity also helped to reduce the thickness of the fouling layer as compared to that under static conditions. Microfiltration membranes were thus utilized for the ultrafiltration of milk. The rapid accumulation of micellar deposits, initiated by adsorption transformed the membrane process into a true ultrafilter or a dynamic membrane capable of effectively retaining serum proteins.

The phenomenon of adsorption was also studied using glass (SiO_2) membranes (Langer and Schnabel 1990). The surface of porous glass contains a large proportion of silanol groups which have a strong tendency to adsorb protein. The effect of protein concentration (e.g. bovine serum albumin) on the amount adsorbed was found to be very significant (0.8 mg/m^2) for unmodified membrane. Surface modification with alcoholic OH groups was shown to reduce the nonspecific binding to a very low level of $2 \mu\text{g/m}^2$ BET surface area. This is believed to be lower as compared to the nonspecific protein binding levels achieved with most ultrafiltration membranes (Langer and Schnabel 1988).

4.3.3. Membrane Surface Characteristics

The surface of a membrane can be modified to enhance or prevent solute interaction with the microporous structure. Membrane surfaces can exist in various forms, i.e. charged, uncharged, hydrophobic or hydrophilic.

Langer and Schanel (1990) studied the use of silica capillary membranes for the separation of proteins. Using glass capillary membranes with hydro-

Figure 4.4. Scanning electron micrograph showing fouling due to Ca and P salts which deposit on the membrane (Vetier, Bennasar and Tarodo de la Fuente 1988). (a: surface of membrane layer showing Ca and P micelles of various sizes; b: cross-section of membrane showing the very thin layer of Ca and P salts.)

phobic characteristics, they showed that a mixture of proteins consisting of lysozyme (MW = 14,400), bovine serum albumin (MW = 68,000), γ-globulin (MW = 150,000) and ferritin (MW = 450,000) can be separated at low applied pressure differences and using high cross-flow velocities. This was achieved by minimizing the interaction of membrane surface with protein molecules through surface modifications converting the silanol groups on unmodified SiO_2 membranes to alcoholic OH groups with low protein affinity. The Stokes radii for the various test molecules varied from about 2 nm (MW = 10,000) to above 10 nm (MW = 1,000,000).

In another study, the zeta potential of the alumina membrane surface was shifted to the positive side (zeta potential decrease) as compared to that of the base membrane by introducing sulfonic groups and to the negative side by amino groups (zeta potential increase). These modified alumina membranes were prepared by reacting the hydroxyl groups with silane coupling agents (Shimizu et al. 1987). There was no change in the surface free energy of the modified alumina membranes as compared to that of the untreated α-alumina membrane.

Surface-modified membranes produced by such techniques may offer certain advantages in biological applications. For example, alumina membranes have a positive zeta potential (+ 5 mV) in a neutral solution and, as a result, proteins with negative potential such as γ-globulin and bovine serum albumin are adsorbed electrostatically. For these systems, the hydraulic resistance increases with a consequent flux decrease. By proper surface modifications it is possible to eliminate or minimize flux decrease in membrane reactors.

The selectivity for the separation of low molecular weight solutes from mixtures can often be enhanced by reducing the membrane pore size. The pore structure within the 40 Å Membralox® transition alumina layer was chemically modified with silanes in an attempt to reduce the pore size for the selective separation of low molecular weight liquid mixtures (Miller and Koros 1990). For the separation of 10 wt.% toluene (MW = 112) from a higher molecular weight lube oil (MW = 660) a maximum selectivity (defined as proportional to the ratio of toluene in permeate to that in feed) of 1.8 was reported. These preliminary data suggest that a nominal reduction in pore diameter may have occurred since untreated 4 nm alumina membranes show no separation capabilities for such liquid mixtures.

4.3.4. Overall Transport Resistance: Determination of Membrane Permeability

The total hydraulic resistance to transport of solvent, solute or particulate retention, as in MF and UF, can be expressed using the simple concept of additive resistances. The permeability of the membrane structure can thus be expressed as the ratio of drivng force (ΔP_T) to the total equivalent resistance to transport:

$$Q_m = \frac{\Delta P_T}{R_m + R_a + R_g + R_b} \tag{4.6}$$

where

$$R_m = \frac{\Delta P_T}{Q_w} \tag{4.7}$$

In practice, however, the various types of transport resistances are often difficult to measure independently. The value of R_m is often obtained from the characterization data using pure liquids showing no adsorption tendency with the membrane material, under the operating conditions. With pure liquids, the transport resistance due to gel layer and concentration polarization phenomenon is generally negligible or absent and thus the flux values vary linearly with ΔP_T. Clean filtered water is often the liquid of choice for the purpose of flux characterization.

Typical water permeability values for a number of commercial inorganic microfiltration and ultrafiltration membranes are given in Tables 4.1 and 4.2. These illustrate the general dependence of water permeability on the nominal

Table 4.1. Typical Water Permeability Values for Commercially Available Inorganic Microfiltration Membranes at 20°C

Manufacturer (Trade Name)	Membrane Material	Pore Size (μm)	Membrane Geometry	Permeability (L/m²-h-bar)
Alcoa/SCT (Membralox®)	α-Al$_2$O$_3$	0.2	Multichannel tubular	2000
Anotec/Alcan (Anopore®)	α-Al$_2$O$_3$	0.2	Plate	3600
Carbon Lorraine	Carbon	0.2	Tubular	1500
Ceramem	SiO$_2$, Al$_2$O$_3$	0.2	Honeycomb	400
Ceram filters	SiC	0.2	Monolith multichannel	
Du Pont (PRD-86)	Cordierite, Mullite, etc.	0.5	Spiral-wound hollow tubules	500
Mott	SS, Ni, etc.	0.5	Tubular	1300
NGK	α-Al$_2$O$_3$	0.2	Multichannel tubular	1500
Norton (Ceraflo®)	α-Al$_2$O$_3$	0.2	Multichannel tubular	2500
Osmonics[†] (Hytrex®)	Ag	0.2	Tubular plate	9000
Pall	SS, Ni, etc.	0.5	Tubular plate	1500
Tech Sep (Carbosep®)	ZrO$_2$	0.14–0.2	Tubular	600

Minneci and Paulson (1988).

Table 4.2. Typical Water Permeability Values for Commercially Available Inorganic Ultrafiltration Membranes at 20°C

Manufacturer (Trade Name)	Membrane Material	Pore Size (nm)	Membrane Geometry	Permeability (L/m²-h-bar)
Alcoa/SCT (Membralox®)	γ-Al₂O₃	4	Multichannel	10
		50	Tubular	300
	ZrO₂	20	Multichannel	400
		50		800
		100		1500
Anotec/Alcan (Anopore®)	γ-Al₂O₃	20	Plate	1000
Carre/Du Pont	Zr(OH)₄–PAA on SS (dynamic)	—	Tubular	
Ceramem	SiO₂, Al₂O₃	50	Honeycomb	250
Gaston County (Ucarsep®)	ZrO₂ on carbon (dynamic)	—	Tubular	
Schott Glass	SiO₂ (glass)	10	Tubular	
Tech Sep (Carbosep®)	ZrO₂	23	Tubular	70
		83		300
TDK (Dynaceram®)	Al₂O₃	50	Tubular	250

pore diameter. In practical situations, the observed flux values at the operating ΔP_T are seldom comparable to the water flux data. This is due to the increased hydraulic resistance to transport across the membrane structure as a result of particle deposits or the formation of gel layer on the membrane.

The resistance due to adsorption may be obtained as discussed in Section 4.3.2, whereas the contribution due to gel layer and the boundary-layer are often less straightforward to obtain independently. In the case of microfiltration, concentration polarization is more apparent as a "cake" on the membrane surface, in contrast to gel polarization which is more common with soluble or finely colloidal substances in applications involving ultrafiltration membranes.

Gel polarization is commonly characterized by the thick layer of rejected material that builds up immediately adjacent to the contacting membrane surface (Figure 4.1). Under gel polarization, the flux is independent of ΔP_T (Figure 4.2), whereas the boundary-layer resistance is dominant up to the point of operation where the flux varies nonlinearly with cross-flow velocity. Here, the system is believed to be in the so-called mass transfer controlled regime. High cross-flow velocities tend to diminish the thickness of the boundary-layer, but may not have any significant influence on the adsorption layer or gel layer.

Thus, by carefully controlling the operating parameters such as the type of liquid, ΔP_T and cross-flow velocity, it may be possible to separate the various transport resistances which is extremely valuable for process optimization. The practical difficulty, however, lies in the fact that often the presence of a thin adsorption layer complicates the quantitative estimation of transport resistance due to gel layer or concentration polarization and hence the determination of effective membrane permeability.

4.4. MICROFILTRATION

Microfiltration membrane modules are typically operated in one of two modes, cross flow or through flow. The discussion on the operating configurations is deferred to Chapter 5 (see Section 5.7). Cross-flow microfiltration is becoming increasingly popular as the preferred mode for a large variety of filtration applications involving separation and concentration of particulate suspensions or solutions, the recovery of low molecular weight substances and in some instances the recovery of macromolecules such as proteins. Although the mathematical models discussed below have been used to explain experimental observations using microporous polymeric membranes, the analysis can be easily extended to describe the transport and separation through microporous inorganic membranes.

4.4.1. Models for the Prediction of Flux

Until recently the observed flux for microfiltration membranes was calculated using the concentration polarization concept developed to predict the ultrafiltration flux. The flux was assumed to be limited by the formation of a concentration polarization boundary-layer. The particles retained by the membrane were thought to be concentrated in this laminar boundary-layer and the value of filtrate flux through the boundary-layer and the membrane wall was obtained by using the following equation (Zydney and Colton 1986):

$$J = \frac{D}{\delta} \ln\left(\frac{C_w}{C_b}\right) = k \ln\left(\frac{C_w}{C_b}\right) \qquad (4.8)$$

The Leveque equation (Leveque 1928), originally developed to predict heat transfer rate through a laminar boundary layer, can be applied to the case of microfiltration. The following equation to calculate the filtrate flux is derived from the model proposed by Leveque:

$$J = 0.807 \left(\frac{D^2 S_w}{\delta}\right)^{1/3} \ln\left(\frac{C_w}{C_b}\right) \qquad (4.9)$$

The above model has been only partially successful in explaining experimental observations and several investigators have reported discrepancies between the predicted values and experimental data (Blatt et al. 1970, Porter 1972, Henry 1972). In other investigations, a model was proposed based on the concept of formation of an immobile cake on the membrane surface at the feed and solid membrane interface that limits the flux. Here the volumetric filtrate flux was given by

$$J_v = \frac{\Delta P_T}{R_c} \tag{4.10}$$

where

$$R_c = \left[\frac{5\mu(1-\varepsilon)^2 \delta_c}{\varepsilon^3} \right] \left(\frac{S_p}{V_p} \right)^2 \tag{4.11}$$

Equation (4.11) is derived using the Kozeny–Karman equation. This model, although in some respects represents an improvement over models derived on the basis of Leveque equation, is also of limited applicability (Porter 1972). Several other models have been proposed to correlate experimental data (Vassilieff, Leonard and Stepner 1985, Madsen 1977, Altena and Belfort 1984).

For most microfiltration applications, the filtrate flux is often several times smaller as compared to the pure solvent flux under otherwise identical conditions. Further, flux decrease is commonly observed during operation. This occurs with increased throughput even under conditions where the particle concentration is relatively constant. This suggests that for practical microfiltration applications, the flux is often controlled by concentration polarization and not by the membrane wall thickness.

This is consistent with experimental observations (Stepner, Vassilieff and Leonard 1985, Zydney and Colton 1986). For membrane-resistance-controlled operations, such as those in some ultrafiltration and reverse osmosis processes, the filtrate flux values are found to increase with increase in the transmembrane pressure (Porter 1972).

A mathematical model that can satisfactorily explain the observed behavior, i.e. flux decrease as a result of particle build-up at the membrane, was proposed. The model envisages correlating the effects of particle migrations in the presence of shear to changes in effective particle diffusivity (Zydney and Colton 1986). The filtrate flux is given by

$$J = 0.078 \left(\frac{r_0^4}{L} \right)^{1/3} S_w \ln \left(\frac{C_w}{C_b} \right) \tag{4.12}$$

The boundary-layer thickness can be calculated using

$$\delta = \frac{D}{k} = 0.578(r_0^2 x)^{1/3} \tag{4.13}$$

The model was tested for various types of feeds containing particulate suspensions including plasmapheresis experiments with blood suspensions where the flux was independent of ΔP_T (gel polarization condition). Experimental data were found to be in good agreement with theoretical predictions using Equations (4.12) and (4.13).

4.5. ULTRAFILTRATION

Ultrafiltration in the conventional sense is a pressure-driven process often used for simultaneously purifying, concentrating and fractionating macromolecules or fine colloidal suspensions without phase change or interphase mass transfer. In this section, the performance characteristics of inorganic membranes produced by sol–gel techniques and those made by anodic oxidation of aluminum to produce porous anodic films will be discussed. The preparation and general characteristics of these types of membranes are covered in detail in Chapters 2 and 3, respectively. The discussion on dynamically formed ultrafiltration and nanofiltration membranes (generally defined as those with a pore diameter of about 1 nm) is deferred to Section 4.6.

4.5.1. Permeation Models

The permeation and separation characteristics of inorganic membranes can be elucidated using the pore model of Ferry (1936). Ferry's model has been utilized to explain the permeation and rejection behavior of monodispersed polyethylene glycols, bovine serum hemoglobin and bovine serum γ-globulin with porous anodic aluminum oxide films (Sugawara et al. 1986). These types of porous inorganic membranes have a well-defined pore structure with straight cylindrical pores and a narrow pore size distribution. The sol–gel synthesized membranes also show a well-defined pore structure although their pore structure is not composed of perfect straight pores (e.g. spherical or slit-shaped pores). The pore model, however, is applicable to sol–gel type membranes as well.

The flux through microporous inorganic membranes with well-defined pore structure can be visualized as a bundle of capillary tubes with uniform pore diameter. The flux can then be expressed by the Hagen–Poiseuille equation:

$$J = \pi r_p^4 (P_1 - P_2)\left(\frac{N}{8\mu t_m}\right) \tag{4.14}$$

Alternatively, the laminar flow through the porous inorganic matrix can be described by the so-called Kozeny–Carman equation (Carman 1937) which can be derived from Equation (4.14).

$$J = \frac{(P_1 - P_2)\varepsilon^3}{\beta \mu t_m S_i^2 (1 - \varepsilon)^2} \tag{4.15}$$

The Kozeny–Carman constant is defined as

$$\beta = f_0 f_t \tag{4.16}$$

Equation (4.15) is analogous to Equation (4.11) used in Section 4.4.1. The numerical values of f_0 and f_t depend on the shape of the pores and on the tortuosity, respectively (Leenaars and Burggraaf 1985a).

The transport rate through ultrafilters may also be obtained through the use of a concentration polarization model. According to the model proposed by Kimura and Sourirajan (1967), the bulk concentration C_b is related to the surface concentration C_s as shown below:

$$J = k \ln \left(\frac{C_s - C_p}{C_b - C_p} \right) \tag{4.17}$$

The value of k can be obtained by using the Leveque equation (Leveque 1928):

$$N_{Sh} = 1.62 \left[N_{Re} N_{Sc} \left(\frac{d}{L} \right) \right]^{1/3} \tag{4.18}$$

4.5.2. Flux Characterization

The linear relationships between flux and transmembrane pressure and between flux and inverse of solvent viscosity for sol–gel ultrafiltration membranes and anodic ultrafiltration membranes are shown in Figures 4.5 and 4.6, respectively. For such microporous membranes complete flow characterization is possible with only a few parameters. This is helpful in tailoring membrane properties to meet application requirements. In the case of polymeric ultrafiltration membranes this situation is analogous to that for the track-etch Nuclepore membranes commonly made of poly (bisphenol-A-carbonate).

Titania and zirconia ultrafiltration membranes were synthesized and characterized by means of water flux measurements (Larbot et al. 1989). Water flux data are reported for titania and zirconia membranes with pore diameters in the range 3–180 nm. These are shown in Table 4.3. It is evident these water flux values are significantly larger than those reported for γ-

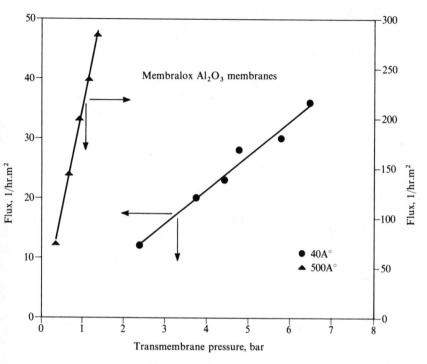

Figure 4.5. Flux vs. transmembrane pressure relationship for sol–gel ultrafiltration membranes (Hsieh, Bhave and Fleming 1988). 40 Å (□) and 500 Å (○) Membralox® alumina membranes.

Table 4.3. Water Flux Data for Developmental TiO₂ and ZrO₂ Membranes (Larbot et al. 1989)

Membrane Material	Pore Diameter (nm)	Transmembrane Pressure (bar)	Flux (L/m²-h)
TiO₂	3	20	4500
TiO₂	20	15	7000
TiO₂	75	10	8000
TiO₂	180	10	11,000
ZrO₂	3	20	2000
ZrO₂	10	15	2500
ZrO₂	20	10	3000
ZrO₂	50	10	5000
ZrO₂	85	10	7000

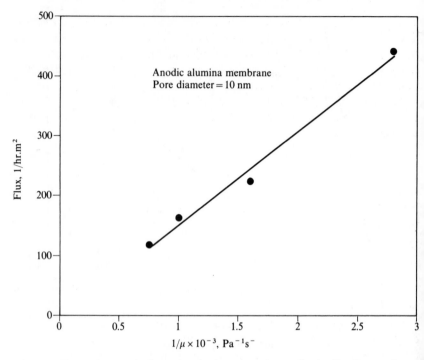

Figure 4.6. Flux vs. inverse of solvent viscosity dependence for anodic ultrafiltration membranes (Itaya et al. 1984). (Pore diameter = 10 nm.)

Al_2O_3 based UF membranes (Larbot et al. 1987, Leenaars and Burggraaf 1985b, Gillot and Garcera 1984, Bhave and Fleming 1988, Hsieh, Bhave and Fleming 1988). These data suggest the presence of defects such as microcracks and other nonuniformities in pore structures (see also Section 2.2.3 under zirconia membranes).

Water permeability values were recently reported for Membralox® zirconia UF membranes of comparable pore diameters (Bhave and Filson 1989, Gillot, Soria and Garcera 1990). For UF membranes at 20°C, the values for 50, 70 and 100 nm pore diameters were reported as 800, 1100 and 1500 m^2 L/h-m^2-bar, respectively. Additional water flux data on inorganic UF membranes are given in Table 4.2.

4.5.3. Solute Retention Properties

The solute properties of inorganic UF membranes have been reported in a number of investigations. A variety of solutes from low molecular weights of about 1000 to as high as 150,000 were evaluated to obtain solute retention data or molecular weight cutoff characteristics. The results of these investigations are reviewed here to provide an understanding of the solute rejection or retention characteristics of the various commercial and developmental inorganic UF membranes.

The solute molecular diameter can be calculated using the Stokes–Einstein equation if data on liquid phase diffusion coefficient and solution viscosity are available.

$$d_s = \frac{KT}{(3\pi\mu D_s)} \tag{4.19}$$

If the value of the pore diameter is known from independent measurements (e.g. gas permeation data) then the ratio of solute diameter, d_s, to pore opening, d_p, can be determined.

$$\alpha = \frac{d_s}{d_p} \tag{4.20}$$

The value of α may be used to indicate the influence of steric factors on the membrane rejection characteristics.

A comparison between the observed and theoretically predicted rejection coefficients is shown in Figure 4.7 as a function of d_s/d_p, where d_p is the membrane pore diameter and d_s is the solute molecular diameter. The rejection coefficient was found to be almost unity as the value of α approached 1. This is only true for perfectly uniform pore diameter membranes such as the anodic alumina membranes.

The knowledge of solute diameter may thus be useful in some situations, to determine a priori, if the membrane of a given pore size can retain certain solutes. However, to maximize solute retention, the value of α should be generally greater than 1. The relative magnitude of α will also depend on the nature and molecular weight of solute.

An interesting study was recently reported for the retention of proteins using symmetric glass membranes with mean pore diameters in the range 12–33 nm (Langer and Schnabel 1990). The proteins tested were lysozyme (MW = 14400, d_s = 3.8 nm), bovine serum albumin (MW = 69,000, d_s = 6.8 nm), immunoglobulin G (MW = 150,000, d_s = 9 nm) and ferritin (MW = 450,000, d_s = 13.5 nm). The retention behavior as a function of membrane pore diameter is shown in Figure 4.8. As might be expected, smaller pore diameter membranes showed larger protein rejection coefficients for high molecular weight proteins such as immunoglobulin (IgG) and ferritin.

The retention of bovine serum albumin (BSA) was almost 100% with a 12 nm pore diameter membrane although the d_s value was only 6.8 nm (and

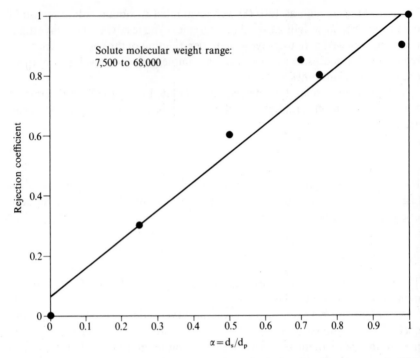

Figure 4.7. Comparison of theoretical and experimental rejection data (Sugawara et al. 1986). ($\alpha = d_s/d_p$, R = rejection coefficient, solute molecular weights ranged from 7500–68,000.)

hence an α value of less than 1). This behavior is typical of proteins such as BSA due to their tendency to form gel-polarized layers during the ultrafiltration process (Matsumoto, Nakao and Kimura 1988). The retention of proteins for a given pore diameter membrane may also vary with the nature of pores and membrane material characteristics. For instance, glass membranes with pore diameter of 20 nm showed a BSA rejection of about 50% (Langer and Schnabel 1990), whereas the retention of BSA on isoporous anodic alumina membranes was found to be negligible for pore diameters in the range 20 nm–0.2 μm (Fane and Hodgson 1990).

The membrane rejection coefficients for anodic alumina membranes were determined using aqueous monodispersed polyethylene glycols of molecular weights 2000, 4000, 7500, 18,000, 39,000 and 150,000 (Itaya et al. 1984). Polystyrene with molecular weights in the range 1000–43,000 and solutions of vitamin B_{12} (MW = 1355), bovine serum hemoglobulin (MW = 64,000), albumin (MW = 67,000) and γ-globulin (MW = 150,000) were also evaluated. The solute concentration in the liquid phase was held constant at 1 g/L at the operating temperature of 20°C.

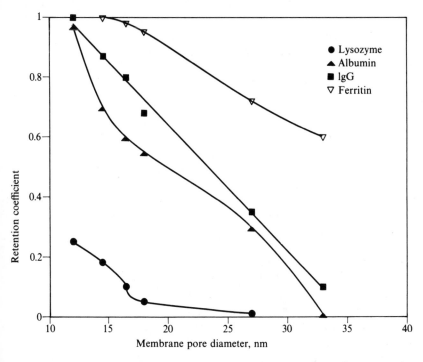

Figure 4.8. Retention behavior as a function of membrane pore diameter for porous glass membranes (Langer and Schnabel 1990).

Figure 4.9 shows the solute retention data for aqueous polyethylene glycols and proteins and for solutions of polystyrene in benzene. These data illustrate the molecular weight cutoff characteristics of anodic alumina membranes. The pore diameter was determined to be 10 nm. The cutoff value for 90% solute rejection was found to be approximately 10,000. This value is about 50% lower than that reported with a similar pore size polymeric ultrafiltration membrane (Baker and Strathmann 1970).

In the above study Itaya et al. (1984) report that no specific decrease in flux values was observed with freshly prepared membranes. This suggests the absence of concentration polarization and also that the solutes tested showed little or no adsorption on to the membrane surface. Such behavior was also independently confirmed by Sugawara et al. (1986). By using dilute solutions (0.1 wt.%) of polyethylene glycols and proteins they were able to minimize concentration polarization effects. Anodic alumina membranes with pore diameter of 10 nm showed a substantially higher solute rejection as compared to that observed with the 18 nm pore diameter membrane. In contrast

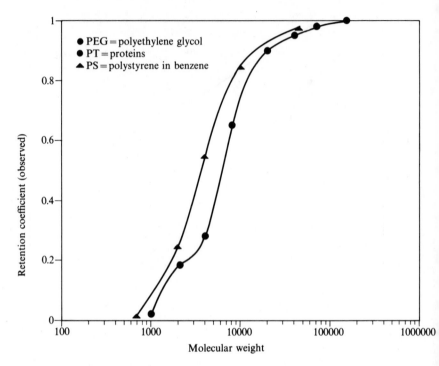

Figure 4.9. Rejection characteristics of anodic UF membranes (Itaya et al. 1984). Aqueous polyethylene glycols and proteins are denoted by (●) and solutions of polystyrene in benzene are denoted by (▲).

many polymeric ultrafiltration membranes (e.g. polysulfone and polypropylene membranes) show a stronger adsorption tendency especially for protein solutions (Bauser et al. 1982, Zeman 1983).

In another developmental effort to prepare and characterize sol–gel ultrafiltration membranes (Leenaars and Burggraaf 1985b) three types of membranes were prepared using sintering temperatures in the range 500–800°C. Some properties of the developmental alumina membranes along with flux and solute retention data are reported in Table 4.4. The solute retention curves are shown in Figure 4.10. The molecular weight cutoff values for Al_2O_3-400 (slit-shaped pores with 2.7 nm width) and Al_2O_3-800 (slit-shaped pores of 4 nm width) as determined from rejection data with PEG were found to be 2,000 and 20,000, respectively. These sharp molecular weight cutoff characteristics for such marginally different pore diameters (viz. 2.7 and 4 nm) further illustrate the relatively narrow pore size distribution of inorganic Al_2O_3 membranes.

Table 4.4. Membrane Properties, Flux and Solute Retention Characteristics of Developmental Al$_2$O$_3$ Membranes (Leenaars and Burggraaf 1985b)

	Al$_2$O$_3$-400*	Al$_2$O$_3$-500*	Al$_2$O$_3$-800†
BET surface area, (m²/g)	334	284	183
Modal slit width, (nm)	2.7	2.9	4
Porosity, %	56	56	55
Water flux (cm³/cm²-h-bar)	0.14	0.15	1.1
Solute retention (%)			
PEG polymers‡			
MW 400			
MW 1000	39		
MW 3000	78		1
MW 6000	94		24
MW 20,000	97		34
MW 100,000			87
			90
			95
Dextrans			
MW 10,500		94	84
MW 70,400			96
MW 465,000		97	97

* $\Delta P_T = 3$ bar;
† $\Delta P_T = 1$ bar;
‡ Molecular weight in daltons

Vycor glass membranes with a pore diameter of about 3 nm show high retention values for low molecular weight solutes such as phenol as compared to alumina membranes of similar pore size (Littman and Guter 1971). This is probably attributable to the differences in membrane surface properties. On the other hand, for higher molecular weight solutes (> 1000) the cutoff values for alumina membranes are in the same range as those obtained with Vycor glass membranes and lower than those reported for Union Carbide/SFEC membranes (Leenaars and Burggraaf 1985b, Bansal 1977, Veyre and Gerster 1984).

In situations where concentration polarization cannot be eliminated, as is the case in most practical applications, the theoretical rejection coefficient, derived from the pore model (Ferry 1936) is given by

$$R_t = \left(\frac{C_s - C_p}{C_s}\right)$$

(4.21)

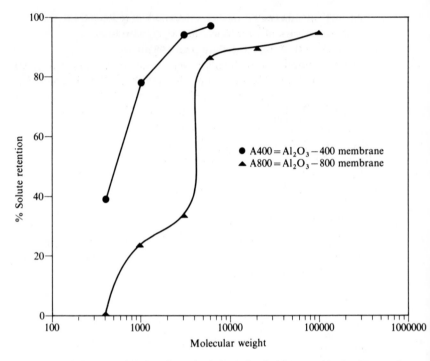

Figure 4.10. Solute retention data for polyethylene glycol polymers with alumina membranes (Leenaars and Burggraaf 1985b). (Al_2O_3-400 and Al_2O_3-800.)

The value of R_t is usually different from the observed rejection coefficient and is given by

$$R_{obs} = \left(\frac{C_b - C_p}{C_b} \right) \tag{4.22}$$

In such situations, the observed rejection coefficient may be substantially lower than the theoretically predicted value. For instance, in the case of nonlinear solutes such as dextrans, it is known that shear deformations take place as the liquid is transported across a microporous membrane (Baker and Strathmann 1970). This effect is more predominant if the chains can move freely as when the pore diameter is large compared to the molecular size of the solute.

The cross-flow filtration of ovalbumin (MW = 45,000) and dextrans (MW = 70,000 and 500,000) using ceramic microfilters with pore diameters of 0.2, 0.8, 1.5 and 3 μm was recently reported (Matsumoto, Nakao and Kimura 1988). The solute concentration was varied in the range 1–10 g/L. It was observed that the rejection of dextrans was very low (only a few percent) whereas the rejection of ovalbumin was greater than 90%. The rejection values for dextrans are consistent with the low molecular size of these molecules. Further, nonlinear molecules such as dextrans undergo shear deformations in cross-flow filtration and do not form micellar colloids. Therefore, such molecules tend to pass relatively freely (compared to linear or less flexible molecules such as PEGs or protein molecules) even when smaller pore size ultrafiltration membranes are used (Baker and Strathmann 1970, Porter 1972). However, the unusually high retention of ovalbumin by the relatively large pore microfilters cannot be explained on the basis of the above listed factors alone.

One of the important contributing factors in the case of proteins such as ovalbumin is the fact that they tend to readily adsorb on the membrane surface (Kimura and Nakao 1984). Thus, in such situations, when cross-flow microfilters are used, severe pore plugging is observed with the formation of "gel polarization" layers that reduce the flux (often dramatically) independent of operating conditions and thus increase solute rejection. Thus separation characteristics of the membrane also strongly depend on the nature of solute.

4.6. LIQUID PERMEATION AND SEPARATION WITH FORMED-IN-PLACE (DYNAMIC) MEMBRANES

In 1965, researchers at the Oak Ridge National Laboratories (Marcinkowsky et al. 1966) involved with desalination research observed that when organic and inorganic polyelectrolytes are added to pressurized salt solutions held in contact with porous supports, salt-filtering layers were formed. These membranes had reverse osmosis properties normally expected of ion exchange membranes. Further these offered moderate salt rejection capabilities with high flux, often an order of magnitude higher as compared to polymeric cellulose acetate membranes. These types of new structures were described as formed-in-place or "dynamically formed" membranes.

During the course of earlier investigations, various kinds of additives such as hydrous oxides, synthetic polyelectrolyte and natural polyelectrolytes (including those found in waste streams and pulp mill wastes) were used to prepare dynamically formed membranes (Kraus, Shor and Johnson 1967, Johnson, Minturn and Wadia 1972, Brandon, Gaddis and Spencer 1981, Groves et al. 1983). These studies concluded that hydrous zirconium oxide $(Zr(OH)_4)$ polyacrylic acid [Zr(IV)–PAA] dual-layer membranes had the best performance characteristics for reverse osmosis applications as compared to many other kinds of formed-in-place membranes (Johnson, Minturn and Wadia 1972).

Many of these formed-in-place membranes [including the Zr(IV)–PAA membrane] do not show high enough salt rejection values for practical desalination applications. However, the formed-in-place membranes with moderate salt rejection characteristics possess the desirable properties for applications requiring higher temperatures especially during processing, cleaning or sterilization. In molecular separations requiring higher operating temperature single-layer dynamically formed inorganic oxide membranes can be utilized instead of the normal two-layer approach involving the second PAA layer (Kimura and Nomura 1981). The PAA layer is also likely to be less temperature resistant and will therefore suffer from poor mechanical stability and operational durability.

The fundamental colloid chemistry underlying the behavior of these membranes is not clear for all cases studied and reported. A few general observations can, however, be made. For instance, in the neutral pH range, the rejection can generally be explained on the basis of polyacrylate layer characteristics. Further, rejection values with single-salt solutions are similar to those observed with single-layer PAA membranes. However, higher rejection values are obtained with dual-layer membranes along with the advantage of higher flux and less sensitivity to support characteristics (Johnson, Minturn and Wadia 1972).

4.6.1. Transport Characterization

The characterization of formed-in-place membranes using transport measurements was reported by Nakao, Nomura and Kimura (1986). The flux, J, was expressed as (Nakao and Kimura, 1982)

$$J = Q_w(\Delta P_T - \sigma \Delta \pi) \tag{4.23}$$

$$R_t = \left[\frac{\sigma(1 - F)}{(1 - \sigma F)} \right] \tag{4.24}$$

and

$$F = \exp\left[\frac{-(1 - \sigma)J_v}{Q} \right] \tag{4.25}$$

Assuming cylindrical pores and using the steric hindrance model (Nakao and Kimura 1982), the observed flux data were explained using

$$J = D_s(1 - \alpha^2)\left(\frac{A_k}{\Delta x} \right) \tag{4.26}$$

where

$$\alpha = \frac{r_s}{r_p} \tag{4.27}$$

Using Equations (4.23)–(4.27), the values of A_k and Δx (membrane thickness) were calculated for the dynamically formed membranes. Dynamic membranes on porous ceramic supports (Nakao, Nomura and Kimura 1986) were prepared by filtering colloid solutions of Zr(IV), Al(III) and Fe(III). Using transport data it was shown that the Zr(IV) dynamic membranes were thinner (lower Δx and higher $A_k/\Delta x$) as compared to Fe(III) membranes. The pore diameters of Zr(IV) and Fe(III) were estimated to be 19 and 25 Å, respectively.

In the absence of gel layer formation on membrane surface or plugging of membrane pores, the permeate flux can be expressed as

$$J = \frac{1}{\mu}\left(\frac{\Delta P_T - \sigma\Delta\pi}{R_m}\right) \qquad (4.28)$$

In these situations, the typical linear relationship between pure water flux and applied transmembrane pressure is analogous to that observed for sol–gel membranes (Figure 4.5) or anodic membranes (Figure 4.6). When gel layers are formed the flux declines with time and may result in increased solute rejection depending on solute concentration and operating conditions. The separation performance under these and various other operating conditions is discussed next.

4.6.2. Separation Performance

The principle of solute separation by formed-in-place membranes, often used in ultrafiltration, nanofiltration or reverse osmosis processes, is based on ion exchange mechanism (Marcinkowsky et al. 1966). It is also known that the most widely studied of all formed-in-place membranes, viz. the Zr(IV)–PAA membranes have the properties of a porous membrane with rejection characteristics that often mimic those normally expected of ultrafiltration or nanofiltration membranes. For example, dynamic membranes formed on porous ceramic supports show good retention properties for substances such as dextrans, PEGs (Kraus, Shor and Johnson 1967) and sucrose (Nomura and Kimura 1980).

Dynamically formed membranes on asymmetric porous ceramic tubes (Dynaceram®, TDK Electronic Co. Japan) were studied for the retention of solutes (Kimura, Ohtani and Watanabe 1985). The ceramic tubes were made of alumina with support pores ranging from 0.5–1.5 μm. The surface (top) layer had a pore size of 0.05 μm (50 nm) with narrow pore size distribution. The solute retention characteristics of dynamically formed ovalbumin ultrafiltration membranes were obtained by filtering through various high molecular weight solutes such as PEGs and dextrans. The observed and theoretical rejection values are shown in Figure 4.11 at two different albumin concentrations. Solute retention data at different pH values in the range 3–9 are shown in Figure 4.12. These data show that very high rejection values were obtained and that the molecular cutoff value of the membrane was about 1000. It may be noted that the theoretical rejection coefficient is a strong function of molecular weight only and was found to be relatively independent of both variations in concentration and pH. On the other hand, for a given molecular weight solute, the observed rejection coefficient varied with both concentration and pH. This suggests that the theoretical calcu-

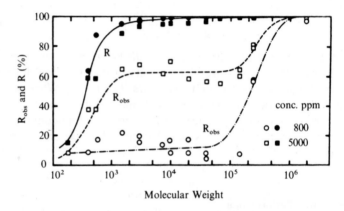

Figure 4.11. Rejection data as a function of solute molecular weight for dynamically formed ovalbumin membranes (Kimura, Ohtani and Watanabe 1985).

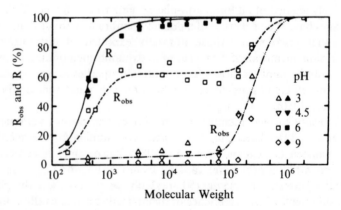

Figure 4.12. Observed and predicted solute retention data for dynamically formed ovalbumin membranes as a function of solution pH (Kimura et al. 1985).

lation of solute rejection, viz. Equation (4.21) or Equation (4.24), should be modified to take into account the effects of variables such as solute concentration and pH. These results suggest that dynamically formed membranes on porous supports may function like small-pore (or low MW cutoff) UF membranes with good retention properties. It may be noted that the separation of dilute liquid mixtures containing dissolved solutes was demonstrated using model compounds such as PEGs, dextrans and proteins. Therefore, care should be exercised in using these results in practical separation applications.

In one practical application, formed-in-place membranes were evaluated for the clarification of diluted molasses of beet and cane sugar (Kishihara, Fujii and Komoto 1986). The ceramic supports used in the study were also asymmetric alumina membranes of 0.05 μm pore diameter supported on a pore support with pore diameters in the range 0.5–1.5 μm. The dynamic membranes evaluated were composed of beet or sugar molasses. The objective of their study was the retention of coloring matter and suspended particles in molasses responsible for turbidity.

Between the beet and sugar molasses formed-in-place dynamic membranes, the beet molasses dynamic membranes showed somewhat lower rejection of suspended particles. This was probably due to the higher porosity of the dynamic membrane layer attributed to the lower amounts of membrane-forming material in beet molasses. On the other hand, the dynamic membranes of cane sugar molasses were denser and of more uniform pore structure, which was probably responsible for the complete retention of particles and significantly higher rejection of coloring matter.

The initial solids content ranged from 10–50 wt.%. It was shown that the retention of suspended matter was almost 100%. For 50 wt.% molasses feed, 2×10^5 Pa pressure and cross-flow velocity of 4 m/s, the filtrate flux at 60°C was reported to be 20 L/h-m^2. The above flux value was about 4 times larger than that obtained with Amicon PM-10 asymmetric polymeric UF membrane. The performance of dynamic membrane for the retention of coloring matter was less satisfactory. Only about 40–60% reduction in coloring matter was achieved as compared with 80% retention with the polymeric UF membrane (MW cutoff value = 10,000).

These results can be explained from the molecular weight cutoff characterization data shown in Figure 4.13. The molecular weight cutoff for the dynamic membrane was determined to be 20,000 at 2×10^6 Pa and 40,000 at 5×10^5 Pa, indicating that the molecular weights of coloring substances were probably substantially lower than 10,000.

In another study, the molecular weight cutoff curves for the Zr(IV), Al(III) and Fe(III) dynamically formed membranes were reported (Nakao, Nomura and Kimura 1986). The results are shown in Figure 4.14. These authors found

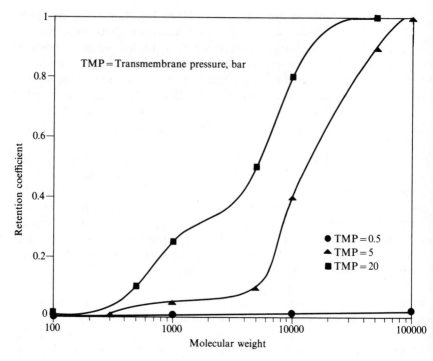

Figure 4.13. Molecular weight cutoff characteristics of formed-in-place dynamic membranes with beet and sugar molasses (Kishihara, Fujii and Komoto 1986).

large differences in rejection values between the Zr(IV) membranes and the Al(III) and Fe(III) membranes. This was due to the fact that very small Zr(IV) colloidal particles formed a layer of lower porosity and with a lower effective pore size. Further, the Zr(IV) particles were small enough to enter the support pore structure in addition to forming a surface layer.

Zr(IV) membranes were also judged to be mechanically superior in contrast to the Al(III) or Fe(III) membranes where the particles were too large to enter the porous structure of the support ceramic tubes and therefore formed a layer on the support surface only. Al(III) or Fe(III) membranes showed poor resistance to mechanical washing as compared to Zr(IV) membranes which could be mechanically washed and showed good durability over a range of applied pressures, temperatures and over a wide pH range. Thus, the overall performance of Zr(IV) formed-in-place membranes was judged to be superior to that of the Al(III) or Fe(III) membranes.

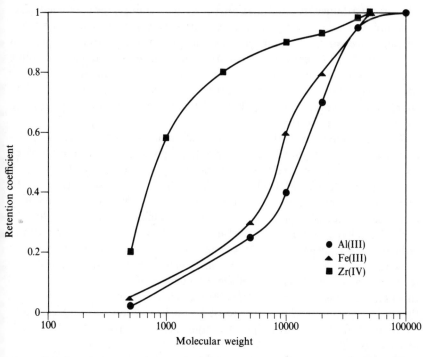

Figure 4.14. Molecular weight cutoff curves for dynamically formed Zr(IV), Al(III) and Fe(III) membranes (Nakao, Nomura and Kimura 1986).

4.7. DENSE MEMBRANES

The permeation of liquids through a dense membrane structure is generally believed to occur by the so-called solution diffusion mechanism. The diffusive flux of each component through the membrane barrier can be expressed by a simplistic model using Fick's law with the assumption that the diffusion coefficient is independent of concentration.

$$J_m = D_s \left(\frac{C_b - C_p}{t_m} \right) \tag{4.29}$$

The effective diffusion coefficient can be represented as a function of temperature and diffusion activation energy (Parlin and Eyring 1954).

$$D_s = D_0 \exp \left(\frac{\Delta E_a}{KT} \right) \tag{4.30}$$

The effective diffusivity value may also be used to indicate the combined effect of solubility and diffusivity of the permeating species in the dense membrane barrier, since solubility values may be difficult to obtain from independent measurements. The effective diffusivity value will reflect the effect of solute concentration on the transport rate through the nonporous membrane barrier. The differences in the effective diffusivity values can be used to characterize the separation capability of the dense membrane under a given set of operating conditions.

Recently, polymers with inorganic backbone structures such as poly-phosphazenes, [poly[bis(2,2,2-trifluoroethoxy)phosphazene], were prepared and characterized for membrane separation applications at higher temperatures up to 350°C (McCaffrey et al. 1987). Flux measurements in the temperature range 25°C–180°C using aqueous 0.1 M methanol, ethanol, isopropanol and phenol were reported. The diffusion flux values for the various aliphatic and aromatic alcohols were, however, found to be similar indicating poor selectivity. At 25°C, the diffusivity values ranged from 0.8 to 1.4×10^{-6} m^2/s. The flux values were also found to be very low primarily due to be extremely low diffusion coefficient values ($\sim 10^{-6}$ of the order of m^2/s) through the dense solvent-cast membrane films. For example, at 25°C, the flux of isopropanol was reported to be 1.52×10^{-6} g-mol m^2-s increasing to 2.65×10^{-4} at 180°C. The diffusivity values increased from 6.75×10^{-7} m^2/s at 25°C to 1.29×10^{-4} m^2/s at 180°C.

Thus, such mixtures are not amenable for separation using dense poly-phosphazene membranes. This can be attributed to the larger values of interstitial sites within the membrane in comparison with the values of molecular radii which ranged from 1.4–2.1 Å.

A new class of membranes described as organo–inorganic composite membranes appears to be attracting some interest. Organic polymers such as polyphosphazenes or heteropolysilanes were impregnated in microporous inorganic supports such as SiO_2, Al_2O_3, ZrO_2 and TiO_2, in an effort to develop membrane structures resembling nanofilters with pore diameters of about 1 nm (Guizard, Larbot and Cot 1990). Preliminary results appear to suggest that such membrane structures may demonstrate molecular cutoff values below 1000. The retention of yellow acid (MW = 749) was found to be almost 100%, whereas the retention of saccharose (MW = 342) was reported to be only 50%, with NaCl (MW = 58.5) retention values in the range 10–20%. Membranes with improved retention characteristics are likely to be developed in the near future.

NOMENCLATURE

A surface area of contact, m^2

A_k ratio of total cross-sectional area to pore area

C_b	solute concentration in the bulk, g-mol/m^3
C_p	solute concentration in the permeate solution, g-mol/m^3
C_g	solute concentration at which gel formation occurs, g-mol/m^3
C_s	solute concentration at the membrane surface, g-mol/m^3
C_w	solute concentration at the wall, g-mol/m^3
d	equivalent hydraulic diameter, Equation (4.18), m
d_p	pore diameter, m
d_s	solute molecular diameter, m
D	solute (or particle) diffusivity, Equations (4.1)–(4.13), m^2/s
D_0	diffusion coefficient constant, Equation (4.30), m^2/s
D_s	solute diffusity, Equations (4.19)–(4.30), m^2/s
ΔE_a	activation energy, J-K/kcal
f_0, f_t	parameters defined by Equation (4.16)
F	permeate flow rate, m^3/s
J	average filtration flux, m^3/m^2-s or m/s
J_m	molar flux of permeating component, g-mol/m^2-s
J_v	volume flux, m^3/s
k	mass transfer coefficient in the boundary-layer adjacent to the membrane surface, m/s
K	Bolzmann constant, J/kcal
L	channel length, m
N	areal pore density, m^{-2}
N_{Re}	Reynolds number
N_{Sc}	Schmidt number
N_{Sh}	Sherwood number
P_1, P_2	upstream and downstream pressures, respectively, Pa pressure difference (feed to retentate), Pa
ΔP_T	transmembrane pressure, Pa
Q	solute permeability, m^3/m^2-h-Pa
Q_m	overall (or effective) membrane permeability, m^3/m^2-s-Pa
Q_w	water permeability, m^3/m^2-h-Pa
r_0	particle radius, m or μm
R_a	resistance due to adsorption, m^2-Pa-s/m^3
R_b	resistance due to the boundary layer, m^2-Pa-s/m^3
R_c	cake resistance Equation (4.11), Pa-s/m^3
R_g	resistance due to gel layer, m^2-Pa-s/m^3
R_m	membrane resistance, m^2-Pa-s/m^3 in Equation (4.6) or m/s in Equation (4.28)
R_{obs}	experimentally observed rejection coefficient
R_t	theoretical membrane rejection coefficient defined by Equation (4.21)
S_i	internal surface area per unit volume of the porous medium, m^{-1}
S_p	particle surface area, m^2

S_w wall shear rate, s^{-1}
t_m membrane layer thickness, m or μm
T absolute temperature, K
V_p particle volume, m^3
x axial coordinate, Equation (4.13), m

Greek

α parameter defined as (d_s/d_p) in Equation (4.20)
β Kozeny–Carman constant, Equation (4.15)
δ boundary-layer thickness, m
δ_c particle cake thickness, m or μm
ε porosity of membrane layer, Equation (4.15) or cake porosity, Equation (4.11)
μ fluid viscosity, Pa-s
$\Delta\pi$ osmotic pressure difference, Pa
σ reflection coefficient

REFERENCES

Altena, F. W. and G. Belfort. 1984. Lateral migration of spherical particles in porous flow channels: application to membrane filtration. *Chem. Eng. Sci.* **39**: 343–355.

Baker, R. W. and H. Strathmann. 1970. Ultrafiltration of macromolecular solutions with high-flux membranes. *J. Appl. Poly. Sci.* **14**: 1197–214.

Bansal, I. K. 1977. Progress in developing membrane systems for treatment of forest products and food processing effluents. *A.I.Ch.E. Symp. Ser.* **73**(166): 144–51.

Bauser, H., H. Chmiel, N. Stroh and E. Walitza. 1982. Interfacial effects with microfiltration membranes. *J. Membrane Sci.* **11**: 321–32.

Bhave, R. R. and H. L. Fleming. 1988. Removal of oily contaminants in wastewater with microporous alumina membranes. *A.I.Ch.E. Symp. Ser.* **84**(261): 19–27.

Bhave, R. R. and J. L. Filson. 1989. Performance characteristics of ceramic membranes in chemical and biochemical separations. Paper read at the 1st Intl. Conf. Inorganic Membranes, 3–7 July 1989, Montpellier.

Blatt, W. F., A. Dravid, A. S. Michaels and L. Nelson. 1970. Solute polarization and cake formation in membrane ultrafiltration: causes, consequences and control techniques. In *Membrane Science and Technology*, ed. J. E. Flinn. Plenum Press, New York.

Brandon, C. A., J. L. Gaddis and G. Spencer. 1981. Recent applications of dynamic membranes. *ACS Symp. Ser.* **154**: 435–53.

Carman, P. C. 1937. Fluid flow through granular beds. *Trans. Inst. Chem. Engr.* **15**: 150–166.

Charpin, J., P. Bergez, F. Valin, H. Barnier, A. Maurel and J. M. Martinet. 1984. Inorganic membranes: preparation, characterization and specific applications. *High Tech Ceramics, Material Science Monographs*, **38** (c): 2211–35.

Fane, A. G. and P. H. Hodgson. 1990. Characterization of and flow through an isoporous inorganic membrane. *Proc. 1st Int. Conf. Inorganic Membranes*, 3–6 July, 501–05, Montpellier.

Ferry, J. D. 1936. Ultrafiltration membranes and ultrafiltration. *Chem. Rev.* **18**: 373–455.

Gillot, J. and D. Garcera. 1984. New ceramic filtering media. In *Filtra 84, Sixth Congress of*

Filtration and Separative Techniques, 2–4 October 1984, 161–72. Societe Francaise de Filtration, Paris.

Gillot, J., M. Soria and D. Garcera. 1990. Recent developments in the Membralox® ceramic membranes. *Proc. 1st Intl. Conf. Inorganic Membranes,* 3–6 July, 379–81, Montpellier.

Groves, G. R., C. A. Buckley, J. M. Cox, A. Kirk C. D. Macmillan and M. J. Simpson. 1983. Dynamic membrane ultrafiltration and hyperfiltration for the treatment of industrial effluents for water reuse. *Desalination* **47**: 305–12.

Guizard, C., A. Larbot and L. Cot. 1990. A new generation of membranes based on organic–inorganic polymers. *Proc. 1st Intl. Conf. Inorganic Membranes,* 3–6 July, 55–64, Montpellier.

Henry, J. D. Jr. 1972. Crossflow filtration. In *Recent Developments in Separation Science,* ed. N. N. Li, Vol. 2, pp. 205–25. Chemical Rubber Co., Ohio.

Hsieh, H. P. 1988. Inorganic membranes. *A.I.Ch.E. Symp. Ser.* **84** (261): 1–18.·

Hsieh, H. P., R. R. Bhave and H. L. Fleming. 1988. Microporous alumina membranes. *J. Membrane Sci.* **39**: 221–41.

Itaya, K., S. Sugawara, K. Arai and S. Saito. 1984. Properties of porous anodic aluminum oxide films as membranes. *J. Chem. Eng. Japan* **17**(5): 514–20.

Johnson, J. S. Jr., R. E. Minturn and P. H. Wadia. 1972. Hyperfiltration. XXI. Dynamically formed hydrous Zr(IV) oxide-polyacrylate membranes. *J. Electroanal. Chem.* **37**: 267–281.

Kimura, S. and S. Sourirajan. 1967. Analysis of data in reverse osmosis with porous cellulose acetate membranes used. *A.I.Ch.E. J* **13**: 497–503.

Kimura, S. and S. Nakao. 1984. Rokagijutsu. In *Filtration Technology,* ed. The Society of Chem. Engrs. p. 130. Maki Shoten, Japan.

Kimura, S. and T. Nomura. 1981. Purification of high temperature water by the reverse osmosis process. *Desalination* **38**: 373–82.

Kimura, S., T. Ohtani and A. Watanabe. 1985. Nature of dynamically formed ultrafiltration membranes. *ACS Symp. Ser.* **281**: 35–46.

Kishihara, S., S. Fujii and M. Komoto. 1986. Clarification of molasses through self-rejecting membrane formed dynamically on porous ceramic tube. Chemical Engineering Papers. *Kagaku Kogaku Ronbunshu* **2**: 205–11.

Kraus, K. A., A. J. Shor and J. S. Johnson 1967. Hyperfiltration Studies. 10. Hyperfiltration with dynamically formed membranes. *Desalination* **2**: 243–266.

Langer, P. and R. Schnabel 1988. Porous glass membranes for downstream processing. *Chem. Biochem. Eng. Q* **2**(4): 242–44.

Langer, P. and R. Schnabel 1990. Porous glass UF membranes in biotechnology. *Proc. 1st Intl. Conf. Inorganic Membranes,* 3–6 July, 249–55. Montpellier.

Larbot, A., J. A. Alary, C. Guizard, L. Cot and J. Gillot. 1987. New inorganic ultrafiltration membranes: preparation and characterization. *Int. J. High Technol. Ceram.* **3**: 143–51.

Larbot, A., J. Fabre, C. Guizard and L. Cot. 1989. New inorganic ultrafiltration membranes. *J. Am. Ceram. Soc.* **72**(2): 257–61.

Leenaars, A. F. M. and A. J. Burggraaf. 1985a. The preparation and characterization of alumina membranes with ultrafine pores. Part 3. The permeability for pure liquids. *J. Membrane Sci.* **24**: 245–60.

Leenaars, A. F. M. and A. J. Burggraaf. 1985b. The preparation and characterization of alumina membranes with ultrafine pores. Part 4. Ultrafiltration and hyperfiltration experiments. *J. Membrane Sci.* **24**: 261–70.

Leveque, M. A. 1928. On the laws of heat transmission by convection (in French). *Ann. Mines* **13**: 201–299.

Littman, F. E. and G. A. Guter. 1971. U. S. Office Saline Water Res. Develop. Progr. Rep. No. 720.

Madsen, R. F. 1977. *Hyperfiltration and Ultrafiltration in Plate and Frame Systems*. Elsevier Science Publishing Co., New York.

Marcinkowsky, A. E., K. A. Kraus, H. O. Phillips, J. S. Johnson Jr. and A. J. Shor. 1966. Hyperfiltration studies. IV. Salt rejection by dynamically formed membranes. *J. Amer. Chem. Soc.* **88**: 5744–48.

Matsumoto, Y., S. Nakao and S. Kimura. 1988. Crossflow filtration of solutions of polymers using ceramic microfiltration. *Int. Chem. Eng.* **28**(4): 677–83.

McCaffrey, R. R., R. E. McAtee, A. E. Grey, C. A. Allen, D. G. Cummings, A. D. Applehans, R. B. Wright and J. G. Jolley. 1987. Inorganic membrane technology. *Sep. Sci. Technol.* **22**(2–3): 873–87.

Miller, J. M. and W. J. Koros. 1990. The formation of a chemically modified γ-alumina microporous membrane. *Sep. Sci. Technol.* 25(13–15): 1257–80.

Minneci, P. A. and D. J. Paulson. 1988. Molecularly bonded metal microfiltration membrane. *J. Membrane Sci.* **39**: 273–83.

Nakao, S. and S. Kimura. 1981. Analysis of solute rejection in ultrafiltration. *J. Chem. Eng. Japan* **14**: 32–37.

Nakao, S. and S. Kimura. 1982. Models of membrane transport phenomena and their applications for ultrafiltration data. *J. Chem. Eng. Japan* **15**: 200–05.

Nakao, S., T. Nomura and S. Kimura. 1986. Formation and characteristics of inorganic dynamic membranes for ultrafiltration. *J. Chem. Eng. Japan* **19**(3): 221–26.

Nomura, T. and S. Kimura. 1980. Properties of dynamically formed membranes. *Desalination* **32**: 57–63.

Parlin, R. and H. Eyring. 1954. *Transport of Membranes*. John Wiley & Sons, Inc., New York.

Porter, M. C. 1972. Concentration polarization in membrane ultrafiltration. *Ind. Eng. Chem. Prod. Res. Develop.* **11**: 234–248.

Shimizu, Y., T. Yazawa, H. Yanigisawa and K. Eguchi. 1987. Surface modification of alumina membranes for membrane bioreactor *J. Ceram. Soc. Inter. Ed.* **95**: 1012–18.

Stepner, T. A., C. S. Vassilieff and E. F. Leonard. 1985. Cell plasma interactions during membrane plasmapheresis. *Clin. Hemorheology* **5**: 15–26.

Sugawara, S., I. Sakurai, M. Konno and S. Saito. 1986. On the permeation mechanism of ultrafiltration with an anodic aluminum oxide membrane. *J. Chem. Eng. Japan* **19**(5): 477–80.

Vassilieff, C. S., E. F. Leonard and T. A. Stepner. 1985. The mechanisms of cell rejection in membrane plasmapheresis. *Clin. Hemorheology* **5**: 7–14.

Vetier, C., M. Bennasar and B. Tarodo de la Feunte. 1988. Study of the fouling of a mineral microfiltration membrane using scanning electron microscopy and physiochemical analyses in the processing of milk. *J. Dairy Research* **55**: 381–400.

Veyre, R. and D. Gerster. 1984. Industrial presence of Carbosep ultrafiltration mineral membranes. Paper read at the 188th American Chemical Society Meeting, 26–31 August, Philadelphia.

Zeman, L. J. 1983. Adsorption effects in rejection of macromolecules by ultrafiltration membranes. *J. Membrane Sci.* **15**: 213–230.

Zydney, A. L. and C. K. Colton. 1986. A concentration polarization model for the filtrate flux in crossflow microfiltration of particulate suspensions. *Chem. Eng. Commun.* **47**: 1–21.

5. Liquid Filtration and Separation with Inorganic Membranes: Operating Considerations and some Aspects of System Design

R. R. BHAVE

Alcoa Separations Technology, Inc., Warrendale, PA

5.1. INTRODUCTION

There are a variety of industrially important liquid phase systems where membrane technology has been successfully applied for separation purposes. Such membrane processes as microfiltration (MF), ultrafiltration (UF) and reverse osmosis (RO) using polymeric membranes have been in commercial use for over two decades.

In recent years, however, inorganic membranes are also being considered in MF and UF applications. A significant number of commercial processes already exist that utilize inorganic membranes. These are discussed in some detail in later chapters (viz. Chapters 8 and 9) dealing with applications. Some of the important inorganic membrane applications are found in the clarification and sterilization of beverages (e.g. wine, beer, fruit juices), cheese making by ultrafiltration, cell harvesting, sterilization of liquids in the pharmaceutical industry, treatment of industrial and municipal water, and wastewater treatment.

5.2. CROSS-FLOW MEMBRANE FILTRATION

A number of experimental and theoretical investigations describing cross-flow filtration using polymeric membranes (Cartwright 1985, Gach 1986, Paulson, Wilson and Spatz 1984, Porter 1986, van Gassel and Ripperger 1985, Zydney and Colton 1986) as well as those for membranes made of inorganic materials such as alumina, zirconia, silica and stainless steel (Galaj et al. 1984, Gillot, Brinkman and Garcera 1984, Norton Company 1984, Matsumoto, Nakao and Kimura, 1988, SCT/Alcoa 1987) can be found in the literature.

Cross-flow filtration, commonly recognized as an efficient fluid management technique, is not of recent origin. This technique was first developed in the early sixties for reverse osmosis and ultrafiltration, where it is used to control concentration polarization. It is only in recent years that cross-flow filtration has been successfully applied to microfiltration.

Cross-flow filtration is a pressure-driven separation process in which the feed stream flows parallel to the filter surface (van Gassel and Ripperger 1985, Gach 1986). Cross-flow filtration helps to improve membrane performance. Unlike through-flow filtration, there are two exiting streams for a given incoming feed stream (Figure 5.1). One exit stream is the permeate (or filtrate); the other is a recirculating stream utilized to provide the sweeping action across the membrane surface. This is needed because one passage does not deplete the feed significantly so that recirculation is required.

Many commercial inorganic membranes, particularly ceramic membranes, have a composite structure with the membrane layer deposited on the inside surface of a tube or channel. The feed flows past the inside surface of the channel and the filtrate flows through the porous support to the shell side of the module. For microfiltration and ultrafiltration, such a parallel flow creates shear forces and/or turbulence to sweep away accumulated particles rejected by the membrane and to prevent or minimize blinding of the filter surface. This is true for all cross-flow membrane modules containing tubular (including multichannel monoliths), spiral-wound hollow tubules and disk-type elements.

Cross-flow filtration is also characterized by the presence of two hydraulic velocities, viz. filtrate velocity perpendicular to the filtering surface (v_1) and

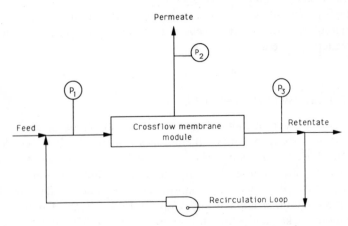

Figure 5.1. Simplified schematic of cross-flow membrane filtration.

cross-flow velocity parallel to the filtering surface (v_2). One of the important objectives in filtration system design is the optimization of the ratio, v_1/v_2. However, extraneous constraints such as capital investment and operating costs often play a major role in the determination of the maximum achievable recovery in terms of the ratio of the filtration rate (R_1) to the recirculation rate (R_2).

Cross-flow membrane filtration processes on an industrial scale, permit (1) concentration of a solute or suspension by reducing the solvent volume in the liquid phase, (2) purification of a liquid by filtering out solutes or particles and (3) fractionation of a mixture of solutes or particles by retaining the desirable ones and letting the others permeate, or vice versa.

5.3. THE EFFECT OF OPERATING PARAMETERS ON MEMBRANE FILTRATION AND SEPARATION PERFORMANCE

There are numerous parameters that may influence filtration performance. These include (1) membrane pore diameter, characteristics of the membrane layer and support structure such as thickness, porosity and wettability characteristics, zeta potential, surface and chemical properties, (2) the nature of the paticulate suspension, which may include the state of agglomeration and/or dispersion, viscosity, charge and presence of dissolved gas and (3) the parameters identifying the filtration conditions such as the pressure, cross-flow velocity, temperature and percentage recovery [per pass (continuous) or total recycle if operated in the batch mode] and backflushing efficiency in the case of microfiltration.

The feed pretreatment, thermal history, pH and other factors can also strongly influence the permeate flux and, therefore, need to be carefully considered at the system design stage. It should be emphasized here that, most often, the prime objective of process optimization is the minimization of fouling since it is the fouling layer that controls the flux.

5.3.1. Membrane Pore Size

Many aspects of the influence of pore size on membrane rejection properties were discussed in Chapter 4 for both microfiltration and ultrafiltration. These are briefly highlighted here to re-emphasize the proper selection of membrane pore size. For ultrafiltration membranes, the value of pore diameter is typically indicative of the solute retention capabilities. In general, the smaller the pore diameter (i.e. the tighter the membrane), the higher will be the rejection coefficient to solutes/particulates of greater nominal size but with a relatively low permeability value (Leenaars and Burggraaf 1985a, b, Hsieh, Bhave and Fleming 1988). Typical clean water permeability values given in

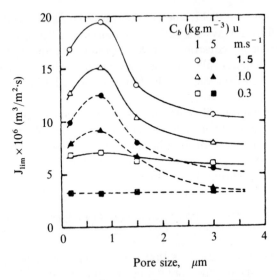

Figure 5.2. Effect of membrane pore diameter on flux (Matsumoto, Nakao and Kimura 1988).

Tables 4.1 and 4.2 (see Chapter 4) illustrate the general dependence of flux on nominal pore size.

In practical applications, the water flux values are seldom comparable to pure water flux values due to a variety of phenomena limiting flux such as adsorption, concentration polarization and gel polarization. On the contrary, in some situations, the optimal pore size for maximum flux may not be related to the permeability of membrane to water, but on the operating conditions that result in the lowest overall transport resistance.

Figure 5.2 illustrates the effect of pore size on flux for ovalbumin filtration (C_b – concentration, u – cross-flow velocity) with alumina microfiltration membranes of pore diameters in the range 0.2–3 μm (Matsumoto, Nakao and Kimura, 1988). It is evident that the pure water flux for the 3 μm membrane is larger than that for the 0.2 μm membrane. However, the limiting flux value for the 0.2 μm pore diameter membrane is higher than that for the 3 μm membrane due to problems associated with severe fouling of larger pore diameter membranes.

5.3.2. Feed Pretreatment

Several pretreatment techniques are employed depending primarily on the characteristics of the feed stream, material corrosion properties and process requirements. The characteristics of the feed stream such as average particle size and particle size distribution, pH, viscosity, etc. are often of prime importance in the development of an effective and efficient treatment procedure. For example, addition of chemicals such as calcium hydroxide, alum and polyelectrolytes to feed streams (e.g. in wastewater treatment) can change the characteristics of the suspended solids and will thus affect the permeate flux. Quite often, the objective of pretreatment is to merely produce a fluffy, cohesionless floc and thereby considerably reduce membrane pore fouling.

In the case of particulate separation applications, the basic purpose of feed pretreatment prior to membrane filtration is to create larger size particles of relatively uniform size and charge density (where desirable) and which can retain their integrity under the operating conditions, including high shear rates (Gramms, Comstock and Hagen 1984, Chen et al. 1991). In other applications such as electrocoat paint recovery and oil–water separations (Bhave and Fleming 1988) it may be necessary to adjust the feed pH to prevent precipitation and/or adsorption of certain undesirable species on the membrane surface or into the pores. In many situations, to avoid channel plugging at the retentate end, the feed is filtered through a screen to remove particles larger than one-tenth (0.1) the size of the channel diameter.

Pretreatment for the filtration of oily wastewater generated in a soybean oil processing plant was described by Bhave and Fleming (1988). Pretreatment of the wastewater with small quantities (10–50 mg/L) of Percol 778 (Allied Colloids, Suffolk, Virginia) as the flocculating agent resulted in the formation of relatively large discrete particles. The permeate, oil and grease, was

Table 5.1. Effect of Feed Pretreatment on Flux: Removal of Oily Contaminants and Suspended Solids from Wastewater and Produced Water

Type of ocess Water	Temperature (°C)	With Pretreatment	Pore Diameter* (µm)	Trans-membrane Pressure (bar)	Flux (L/m²-h) Without Pretreatment	Flux (L/m²-h) With Pretreatment
e oil plant	50	—	0.004	5.1	35	—
stewater† etable	50	—	0.05	2.4	87	
il plant	60	Percol 78:50 ppm	0.2			430
stewater† duced	60		0.2	0.7	27	
ater‡	30	—	0.5	3	< 200	
	30	80–100 ppm of chemical flocculants	0.5	0.9		2000

04 and 0.05 µm γ-Al$_2$O$_3$ Membralox®; 0.2 and 0.5 µm α-Al$_2$O$_3$ Membralox®
ave and Fleming (1988)
en et al. (1991)

reduced to 5–10 mg/L with a COD value in the range 1000–2000 (associated with the small quantities of organic impurities) which was substantially lower than the initial feed value of 10,000. Thus, it was shown that 0.2 μm Membralox® alumina microfilters can be used to perform the filtration rather than conventional ultrafilters, with very substantial gain in flux. Recently, Chen et al. (1991) have also reported the use of feed pretreatment for the filtration of produced water to obtain high flux values (up to 2000 L/h-m^2) using microfilters. Table 5.1 summarizes the results for oily wastewater and produced-water filtrations using ceramic membrane filters.

Unfortunately, it is not always possible to pretreat or to modify the chemistry of the feed to the filtration system. As a result, in most situations, it is necessary to define the smallest particle size present in the process feed (e.g. by sonicating the sample). Using this information, a membrane pore diameter may be chosen which is smaller than the smallest particle.

5.3.3. Cross-flow Velocity

Cross-flow velocity is the average rate at which the process fluid flows parallel to and across the membrane surface. Due to the continuous removal of permeate through the membrane, the cross-flow velocity is somewhat higher at the inlet as compared to that calculated based on the retentate flow rate (depending on the fraction of the feed flowing out as permeate per pass). For many filtration applications (e.g. MF and UF), it is generally recommended to be in the range 2–8 m/s depending on process stream properties such as viscosity and particle loading and constraints imposed by the allowable pressure drop (Cheryan 1986). For instance, in microfiltration applications with alumina membranes, a cross-flow velocity in the range 3–7 m/s is generally employed.

An increase in cross-flow velocity usually results in an increase in flux (Paulson, Wilson and Spatz 1984, Galaj et al. 1984, Bhave and Fleming 1988). This is a consequence of higher shear rates which promote the efficient removal of particles, solutes or of components of the dynamic layer deposited on the membrane surface.

A higher cross-flow velocity, however, causes a substantially higher tube-side pressure drop which may pose a problem to the membrane structure in the case of some polymeric geometries as in hollow fibers. Some tubular systems can operate at somewhat higher pressures. Inorganic membranes, on the other hand, do not suffer from such a limitation and usually have much higher pressure ratings. For example, Membralox® alumina membranes can be operated at pressures in excess of 30 bar (Bhave, Gillot and Liu 1989).

In membrane systems using centrifugal circulating pumps the increase in concentration must be managed carefully to avoid a sudden increase in retentate viscosity and complete blocking of the inside part of the loop. If such a problem occurs, modules must be backflushed with clean deionized water to remove concentrate from the channels before starting the cleaning cycle.

5.3.4. Transmembrane Pressure

For many membrane separations, such as microfiltration and ultrafiltration, the permeate flow rate depends directly on the applied transmembrane pressure for a given surface area under otherwise uniform conditions. The following equation is commonly utilized to calculate the transmembrane pressure (ΔP_T):

$$\Delta P_T = \frac{P_1 + P_3}{2} - P_2 \tag{5.1}$$

The location of P_1, P_2 and P_3 is shown in Figure 5.1.

However, in situations where P_1 is substantially larger (e.g. by a factor of 3 or more) than P_3, a logarithmic-mean-based ΔP_T should be used. The general dependence of flux on ΔP_T is illustrated in Figure 5.3. As can be seen from the figure, at low ΔP_T values the flux increases linearly with ΔP_T up to a point commonly described as the "threshold pressure" (curve b). This is also characterized as the pressure-controlled regime. Beyond the threshold pressure, the flux increases nonlinearly with ΔP_T due to concentration polarization (Porter 1986, Lee, Aurelle and Roques 1984), and at higher ΔP_T the flux

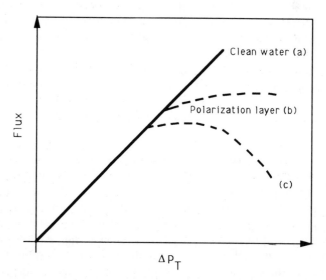

Figure 5.3. General dependence of flux on transmembrane pressure.

becomes independent of ΔP_T due to gel polarization (curve b) which can even cause a flux decline (curve c) if the pressure drop becomes excessive. Such a flow is characterized as belonging to the mass-transfer-controlled regime (Bhave and Fleming 1988).

5.3.5. Temperature

The effect of temperature on flux can be quite significant even within the constraints imposed by the system such as the nature of feed or product. In general, flux increases with temperature due to the decrease in viscosity and/or due to the increase in solubility of suspended solids with an increase in temperature. For most dilute solutions, under otherwise uniform conditions, the effect of temperature on flux can be predicted from a known relationship of viscosity with temperature such as the Stokes–Einstein relation. For example, in the case of dilute aqueous solutions, the flux can be increased by a factor of 2 by raising the temperature from 20–50°C (Bhave and Fleming 1988).

A higher value of flux will reduce the membrane area requirements and can result in a lower overall cost. In some situations, the increased energy costs associated with the higher operating temperature have to be considered to achieve an optimum balance between the increased operating costs and lower capital costs.

The above analysis is valid when the filtration is controlled either by membrane resistance or boundary-layer resistance. Such situations commonly occur when small-pore ultrafilters are used (Bhave and Fleming 1988). However, when concentration polarization is controlling, the effect of temperature on flux will depend on the relationship between the liquid phase mass transfer coefficient and viscosity. In these situations, as in many microfiltration applications (Zydney and Colton 1986), the viscosity dependence of flux can be strongly nonlinear.

5.3.6. pH

The pH value can have a significant influence on the permeation rate especially around the isoelectric point of certain colloids where they tend to destabilize and precipitate. It may also be noted that many inorganic membranes such as Al_2O_3, ZrO_2, SiO_2 and TiO_2 consists of charged particles whose zeta potential values are influenced by solution pH. For example, the isoelectric points of Al_2O_3, ZrO_2, SiO_2 and TiO_2 are known to be around pH values of 9, 5, 2 and 6, respectively. These changes can alter the membrane resistance and hence the flux characteristics of the filtration process. As a result of the interactions between the membrane surface and solutions filtered (such as by adsorption), changes in zeta potentials occur.

The effect of pH on flux and zeta potential for cross-flow filtration of mineral slurries such as SiO_2 using 0.2 μm alumina membranes was recently reported (Hoogland, Fane and Fell 1990). The results are shown in Figure 5.4. The solids loading was 20 g SiO_2 particles per liter of slurry volume. The mean particle size was about 2.4 μm with a significant portion of fines in the range 0.2–1 μm. These data indicate that higher flux characteristics were obtained both at low and high values of zeta potential.

A higher value of zeta potential (or at low pH around 2) resulted in a higher flux value due to the depolarization effect which promotes the formation of a thinner layer of rejected particles on the membrane surface. The depolarization phenomenon is believed to occur at the isoelectric pH which was found to be about 2. On the other hand, higher flux values were also realized at low zeta potentials (or at pH > 9) due to particle agglomeration or flocculation, especially with the larger micron sized SiO_2 particles. This argument for the

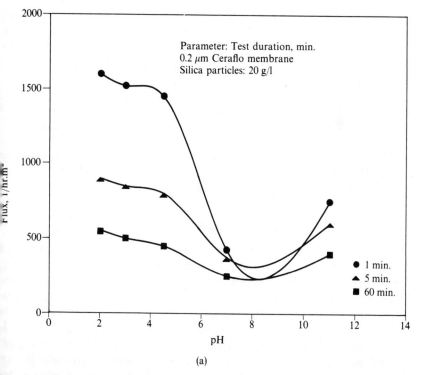

(a)

Figure 5.4. Effect of pH on flux and zeta potential for cross-flow filtration with mineral slurries (Hoogland, Fane and Fell 1990).

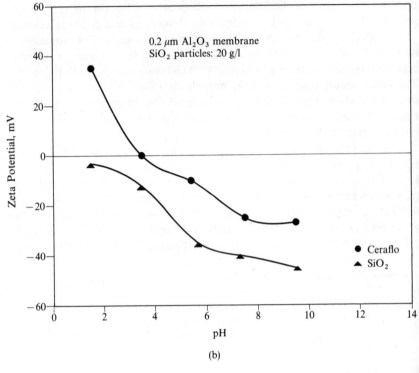

(b)

Figure 5.4 (*continued*)

observed higher flux at pH > 9 suggests that SiO_2 particles can be flocculated at pH 9 which seems incompatible with the isoelectric point (pH = 2) of SiO_2. Further work is necessary to explain these preliminary results.

5.4. BACKFLUSHING: THEORETICAL ASPECTS

During the filtration process, flux decreases as a result of the accumulation of rejected particles on the membrane surface. Cross-flow filtration is often helpful in slowing down membrane fouling but does not eliminate it. Backflushing can help remove the excessive particle deposits on the membrane surface and also minimizes the decrease in flux. It involves the application of periodic counterpressure on the permeate side of the membrane to push a small quantity of permeate through the support into the feed of the module

A mathematical model of membrane surface fouling by particles and of cleaning (or membrane regeneration) by backflushing was developed by Galaj et al. (1984). The model applies to a suspension of particles of uniform

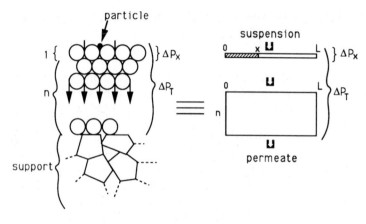

Figure 5.5. A schematic of the filtration model with backflush (Galaj et al. 1984).

...ize in a homogeneous liquid. It assumes that fouling is due to particles being forced into the pore entrances at the surface of the membrane and that the particles can be expelled by backflushing. This requires that the force exerted on the particles by the backflush be equal to or higher than the force pushing the particle into the pore entrance.

Figure 5.5 shows a section of membrane tube of length L filtering a particle suspension under a total system transmembrane pressure, ΔP_T. The main equations which describe the model are summarized below. For full details on the model development the reader is referred to the work of Galaj et al. (1984).

According to Galaj et al. (1984), the value of filtrate flux, F_α, in terms of the fraction of pores plugged is expressed by

$$F_\alpha = F_0 \left[1 - \frac{\alpha}{n(1 - \alpha) + 1} \right] \tag{5.2}$$

The minimum retention ratio, α', for a given plugging condition, α, is given by

$$\alpha' = \frac{\Delta P_B}{\Delta P_T} \alpha + \frac{n + 1}{n} \left[1 - \frac{\Delta P_B}{\Delta P_T} \right] \tag{5.3}$$

Total membrane regeneration is obtained when $\alpha' = 0$. Substituting this value of α' in the above equation,

$$\frac{\Delta P_B}{\Delta P_T} = \frac{n + 1}{n(1 - \alpha) + 1} \tag{5.4}$$

If, however, complete membrane pore blockage occurs, $\alpha = 1$ and

$$\Delta P_B = \Delta P_T(n + 1) \tag{5.5}$$

This shows that under the condition of complete pore plugging the operating pressure differential required for backflushing will be very large. The main conclusions from the above model are summarized below:

1. As the number of blocked pore entrances increases, that is, as the fouling progresses, the flux decreases at first very slowly and then rapidly.
2. An efficient backflushing pulse is essentially characterized by the maximum value of the backflushing pressure. A short pulse with a high pressure is preferable to a long pulse with a low pressure.
3. The backflushing pressure required to dislodge all particles increases with the proportion of blocked pore entrances. This rate of increase is at first very slow and becomes very rapid as the proportion of blocked pores approaches 100%.

The practical implications of the model along with some operational aspects of backflushing are described in Section 5.8.5.

5.5. MEMBRANE REGENERATION

All membrane-based filtration processes such as MF, UF and RO require some form of periodic cleaning to remove the foulants generally by chemical attack or by dissolution. Membrane cleaning may involve the removal of external foulants deposited on the membrane surface as well as foulants embedded within the microporous structure (Cheryan 1986, Chong, Jelen and Wong 1985, Gillot, Brinkman and Garcera 1984, Leenaars and Burggraaf 1985a, b, Norton Company 1984, SCT/Alcoa 1987).

Inorganic membrane materials are chemically inert in strongly alkaline or basic solutions and are not attacked by organic solvents or oxidizing agents even at high temperatures. This allows the development of an effective membrane regeneration procedure with minimal constraints. Inorganic membrane elements can also be steam sterilized or autoclaved (Gillot, Brinkman and Garcera 1984, Norton Company 1984).

It is impossible to formulate a single cleaning procedure to cover the wide range of inorganic membrane applications. In most situations, a collaboration between end user, an engineering company and the cleaning agent manufacturer is the best approach.

A number of cleaners are used in industrial practice depending on the type of fouling encountered. For example, with mineral foulants acid cleaning is generally effective, whereas for oily or soluble organic foulants alkaline cleaning agents are often recommended. In some situations, such as those involving biological debris or materials, enzyme cleaners can serve as alternatives for acid–base cleaning agents.

For severely fouled membranes (e.g. both internally and externally fouled membrane elements), sequential use of strongly acidic and basic solutions may be necessary, including addition of bleaching agents such as sodium hypochlorite. The addition of complexing agents, wetting agents and/or surfactants also quite often greatly improves the cleaning efficiency.

To get the best results from the operational view point, it is generally recommended that a low transmembrane pressure be maintained during the cleaning cycle. Permeate port(s) must be kept closed during the first half of the cleaning duration and kept open during the second half of the cleaning cycle. The use of separate loops for the recirculation of cleaning solutions and rinse water is recommended, especially when some sections or components of the entire filtration system are not resistant to corrosion by the cleaning agents.

The cleaning solutions must be free of dissolved air or gas as this will promote membrane fouling. This is due to the fact that for microporous membranes, a positive pressure is required to dislodge the bubbles or entrapped gas, the value of which increases with decreasing membrane pore diameter. The quality of water used for cleaning and rinsing is also of great importance. Cleaning efficiency is best when the water has a low fouling index, a low hardness and a low concentration in silica, iron and manganese.

5.6. MICROFILTRATION WITH UNIFORM TRANSMEMBRANE PRESSURE

Alfa-Laval have developed an interesting optimization technique for micro-filtration applications (Sandblom 1978). Initially developed for the filtration of milk and dairy products, Alfa-Laval have demonstrated that micro-filtration with uniform transmembrane pressure (also known as the "Bac-tocatch" process) offers the possibility of obtaining a considerably higher permeate flux with > 99.5% bacteria retention (Malmberg and Holm 1988).

In conventional cross-flow microfiltration the transmembrane pressure varies substantially along the length of the module from the inlet to the outlet. This is due to the fact that in conventional cross-flow microfiltration the permeate (filtrate)-side pressure is maintained at a constant value, whereas the feed-side pressure varies from inlet to outlet as a result of the hydraulic pressure loss at a given cross-flow velocity. Typically, cross-flow velocities are high (in the range 3–8 m/s). Thus there is a substantial loss of feed-side pressure as the often viscous mass is pumped through the feed channels. Typical pressure profiles observed in conventional cross-flow microfiltration

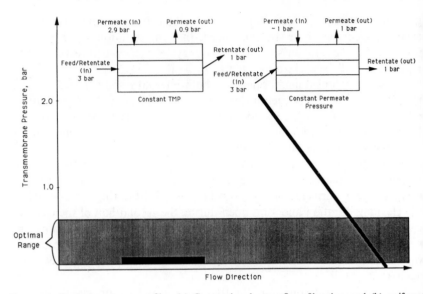

Figure 5.6. Typical pressure profiles: (a) Conventional cross-flow filtration and (b) uniform transmembrane pressure microfiltration (Malmberg and Holm 1988).

and in uniform transmembrane pressure microfiltration are shown in Figure 5.6.

The essential difference between the two operational modes is that the transmembrane pressure in the Bactocatch process is uniform along the entire length of the filtration module, whereas in the conventional cross-flow microfiltration a substantial variation in the transmembrane pressure along the length of the module occurs and is not compensated by external means as is the case with the Bactocatch process. Thus, in the Bactocatch process the permeate-side pressure is not held constant but is varied along the length to adjust the transmembrane pressure to a constant value. This can be readily accomplished with the help of a dynamic counterpressure recirculation pump placed in the permeate-side loop. A schematic of the Bactocatch process is shown in Figure 5.7.

Utilizing such an operational mode, higher flux values are generated (> 500 L/h-m^2 in the case of milk) primarily as a result of high recirculation velocities that help in minimizing fouling due to concentration polarization. Additionally, under conditions of constant transmembrane pressure the useful operational time between successful cleaning cycles is significantly increased. This may be an important advantage in many batch processing applications as it offers higher overall throughput or yield of the desired product.

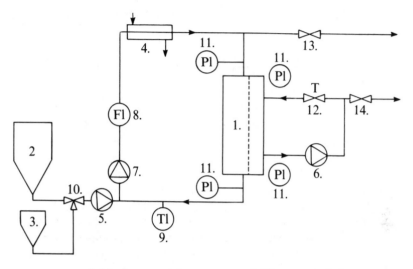

1. Membrane module
2. Feed tank
3. Cleaning tank
4. Plate heat exchanger
5. Feed pump
6. Circulation pump (permeate)
7. Circulation pump (retentate)

8. Electromagnetic flow meter
9. Thermo-element
10. Three-way valve
11. Pressure gauges
12. Regulation valve
13. Retentate outlet valve
14. Permeate outlet valve

Figure 5.7. A schematic of the Bactocatch process for microfiltration with uniform trans-membrane pressure (Tragardh and Wahlgren 1990).

In conventional cross-flow filtration, the flux at a section close to the inlet of the module is higher as a result of higher transmembrane pressure but decreases significantly along the length due to pressure drop at high cross-flow velocities. This promotes the formation of a gel layer of nonuniform thickness along the length (probably thicker at the channel entrance and somewhat thinner at the exit) and can severely reduce the overall permeate flux.

Membralox® microfiltration modules have been utilized to demonstrate this concept in a number of food, dairy and beverage processing applications. Until the end of 1989, approximately 50 pilot-scale and industrial-scale systems (with membrane areas ranging from 1–10 m^2) were sold mostly in Europe (Maubois 1990, Gillot 1990). This type of filtration is performed without backflushing and generally at low transmembrane pressures to obtain an optimum flux that can be sustained over long periods of operation. Flux values in the range 300–1000 L/h-m^2 have been reported (Maubois 1990, Gillot, Soria and Garcera 1990). This concept can be extended to other application areas such as those in biotechnology and the pharmaceutical industry.

5.7. OPERATING CONFIGURATIONS

For the proper operation of a membrane system, a number of requirements need to be satisfied apart from adequate heat transfer to maintain the desired operating temperature and any feed pretreatment, if necessary.

It is essential to maintain a sufficient feed rate to ensure that the desired values of bleed flow rate and permeate flow rate are obtained. It is also necessary to ensure that the desired transmembrane pressure can be maintained which will be influenced by the pressure loss along the module length. Feed-side pressure drop will depend on cross-flow velocity and retentate viscosity. Depending on the viscosity value, either a centrifugal or a positive displacement pump will be required. Three basic operational configurations can be identified. These are: (1) open system (2) closed system and (3) feed and bleed.

5.7.1. Open System

This type of operation is generally preferred for preliminary feasibility tests. A schematic of the system is shown in Figure 5.8. Here, a single pump provides both the needed cross-flow velocity and transmembrane pressure. The disadvantage of such a scheme is the energy loss associated in raising the

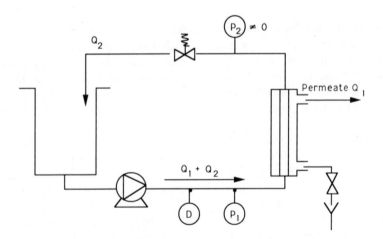

Figure 5.8. A schematic of the operation in the "open system" configuration (D: flowmeter; P_1, P_2: pressures; Q_1: permeate; Q_2: retentate).

pressure from atmospheric to the inlet feed pressure value each time the retentate is recirculated. In addition, the feed characteristics may be adversely affected due to the repeated recirculation action involving flow through the pump and associated valving arrangements.

5.7.2. Closed System

This type of operational mode is illustrated in Figure 5.9. The desired values of cross-flow velocity and transmembrane pressure are maintained by a set of two pumps. A low flow rate feed pump is generally selected with a high-flow recirculation pump to reduce or eliminate the energy loss disadvantage experienced in the open system. However, a negative aspect of such a scheme is the rapid concentration effect in the recirculation loop, which can accelerate fouling and thus reduce the filtrate flux. The actual value of the volumetric concentration factor will depend on the system dead volume, permeate flux and duration of operation.

$$V_c = \frac{V_0 + Q_1 t}{V_0} \qquad (5.6)$$

5.7.3. Feed and Bleed

In this mode of operation, three variations are possible: (1) batch (2) continuous and (3) continuous cascade. These are schematically shown in Figures 5.10, 5.11 and 5.12, respectively.

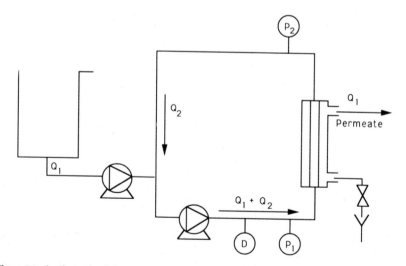

Figure 5.9. A schematic of the operation in the "closed system" configuration (D: flowmeter; P_1, P_2: pressures; Q_1: permeate; Q_2: retentate).

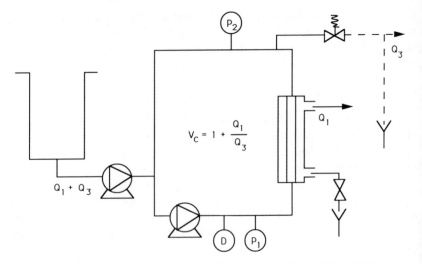

Figure 5.10. A schematic of the operation in the "batch" mode of feed and bleed (D: flowmeter; P_1, P_2: pressures; Q_1: permeate; Q_3: retentate).

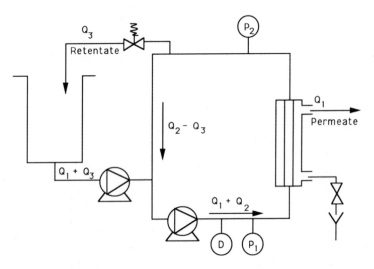

Figure 5.11. A schematic of the operation in the "continuous" mode of feed and bleed (D: flowmeter; P_1, P_2: pressures; Q_1: permeate; Q_2: feed; Q_3: retentate).

1. In the case of a batch operation, the return of a portion of retentate back into the feed tank reduces the rapid concentration effect illustrated above. The actual reduction will depend on the ratio, Q_3/Q_1. Product sensitivity considerations such as thermal or bacteriological may influence the batch duration.

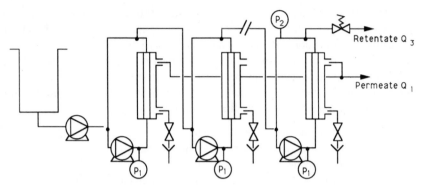

Figure 5.12. A schematic of the operation in the "continuous cascade" configuration of feed and bleed (P_1, P_2: pressures; Q_1: permeate; Q_3: retentate).

2. In the case of a continuous operation, a portion of retentate (Q_3) is permanently bled out of the filtration loop. The volumetric concentration factor in this situation is given by

$$V_c = 1 + \frac{Q_1}{Q_3} \qquad (5.7)$$

The desired value of V_c may be obtained by manipulating operating conditions and permeate flux.

3. When high concentration factors are desired, it may be necessary to place several recirculation loops in series as shown in Figure 5.12. In this case, the desired transmembrane pressure can be obtained by adjusting the retentate pressure on the last loop for a given inlet feed pressure. The total permeate flow rate in a cascade operation, in many situations, can be expressed as a logarithmic function of the volumetric concentration factor:

$$Q = \alpha_1 - \alpha_2 \ln V_c \qquad (5.8)$$

α_1 and α_2 are experimentally determined. As can be seen, the higher the number of stages in a cascade the closer its average flow rate to that of the batch operation. Under this configuration a higher overall flux per unit surface area can be achieved.

5.8. SOME ASPECTS OF SYSTEM DESIGN AND OPERATION

The design and operation of a pilot-scale or larger-scale filtration system for a given application is often based on the qualitative and quantitative information obtained from the operation of a bench-scale or laboratory-scale filtration system.

A variety of information is needed to qualitatively assess the filtration performance. For example, for a given operational mode such as batch or continuous (single stage or multistage), the knowledge of the specific retentions of the components from a solution, emulsion or suspension enables the calculation of the volumetric concentration factor. The analysis of feed, retentate and permeate must be separately established. Several important parameters along with their qualitative interrelationships are given below:

1. Values of membrane pore diameter
2. Physicochemical interaction between the membrane and the various components to be separated or filtered
3. pH which also can influence the physicochemical interaction
4. Temperature which can influence the physicochemical interaction but may also influence other factors such as the microorganism survival or the physicochemical condition of colloids
5. Effect of adsorption and/or polarization phenomena on membrane permeability
6. Operating pressure which can influence the adsorption and/or polarization phenomena
7. Shear rate (or cross-flow velocity) which can influence the adsorption and /or polarization phenomena but may also influence microorganism survival and particle structural integrity
8. Volumetric concentration factor which can influence both the physicochemical interaction and adsorption/polarization phenomena
9. Duration of operation which can modify a variety of factors given under (2), (3), (4), (5) and (8).

The assessment of the relative importance of these various parameters may require the operator to perform many tests in order to make a large number of systematic variations in parameter values. Among other qualitative aspects to be considered include the system "dead volume" which is an important consideration for applications involving bacteria or where physicochemical equilibrium considerations are important.

For a given application and operating constraints, the establishment of certain qualitative data as discussed above is often needed prior to the determination of conditions for flux optimization. These are briefly discussed below.

5.8.1. Transmembrane Pressure

For clean water, the permeate flow rate is proportional to the transmembrane pressure. However, in practical filtration situations, the presence of a concentration polarization layer and/or gel layer limits the flux increase with ΔP_T

and may even cause a decrease in flux with an increase in pressure (as illustrated in Figure 5.3). In some cases, it also influences the solute retention properties of the membrane.

5.8.2. Shear Rate

The shear rate, which commonly depends on the cross-flow velocity, is commonly produced by the displacement of liquid to be filtered by the recirculation pump. The choice of pump such as centrifugal or positive displacement is generally dictated by the value of retentate viscosity. Centrifugal pumps are inefficient for high-viscosity liquids, whereas positive displacement pumps are less efficient to handle low-viscosity liquids. Feed pumps are usually rated for lower flow rates than recirculation pumps but are capable of operating at high pressures.

The shear rate strongly influences the concentration polarization layer. Depending on the prevalent hydrodynamic condition such as laminar flow or turbulent flow, the value of boundary-layer thickness can be estimated for a given recirculation rate (see Section 4.4). Increasing the shear rate generally increases the permeate flux (within certain limits) but at the expense of higher energy consumption. This also results in increased cross-flow pressure drop and may thus affect the transmembrane pressure in conventional cross-flow filtration with inorganic membranes.

In the case of thixotropic liquids such as milk, an increase in shear rate decreases the shear stress and thus decreases the apparent viscosity. This can result in an increase in the permeate flux in certain situations. The relation between shear stress and shear rate can be written as

$$\tau = A \left(\frac{dU}{dr} \right)^a \tag{5.9}$$

This type of information can be very useful in the optimization of the important process variables, viz. cross-flow velocity, viscosity and ΔP_T.

5.8.3. Temperature

The operating temperature can have a positive influence on permeate flux because, in general, an increase in temperature decreases the viscosity and increases the value of diffusion coefficient. This can help reduce the polarization effect which tends to decrease the permeate flux. On the other hand, a change in temperature can alter certain physicochemical equilibria such as denaturation of certain proteins. Thus, an optimum temperature value is determined by considering such aspects and other operating constraints.

Depending on the mode of operation, a tubular, double-wall or coil heat exchanger is utilized to obtain and regulate the desired system temperature.

5.8.4. System Dead Volume

For continuous processes, the lowest possible dead volume will enable the operation with low average holding times, which may be important in some applications such as those involving bacteria laden liquids. For batch processes, the value of dead volume in relation to the initial volume will determine the maximum volumetric concentration factor achievable. It is generally desirable to keep the system dead volume as low as possible.

5.8.5. Backflushing: Operational Aspects

Backflushing is employed in most cross-flow microfiltration and in some ultrafiltration systems as an effective technique to disrupt or destroy the gel layer or concentration polarization layer. It also helps in preventing particle penetration into the porous microstructure which will have a detrimental effect on the flux.

Backflushing is achieved by applying periodic counterpressure on the permeate side of the membrane, often with the help of an automatic time switch or a microprocessor to push back a given permeate volume in the opposite direction. It may be possible to work with lower countercurrent pressures by reducing the retentate pressure, especially in smaller systems. However, when several modules are connected in series, reduction in retentate pressure is not a desirable option due to substantial loss of efficiency. In such a case, the backflushing operation may be performed sequentially on each module.

In Section 5.4, some theoretical aspects of backflushing were described using the model developed by Galaj et al. (1984). The practical implications of the model when applied to a cross-flow filtration operation suggests that (1) frequent, short, high-pressure pulses are preferred and (2) backflushing should start at the very beginning of the filtration operation. In other words, it is not desirable to wait until the flux decreases noticeably to begin the application of backflush. A delay in the application of backflush will require a higher backflushing pressure to clean the membrane.

In situations where no backflushing is used the flux may reduce to virtually zero within minutes. Thus high-efficiency sequential backflushing is recommended to achieve higher flux values and allow the filtration cycle to continue over longer periods of time. There is, however, a certain percentage loss of permeate associated with backflushing due to countercurrent permeation. The increase in flux due to sequential backflushing typically far exceeds

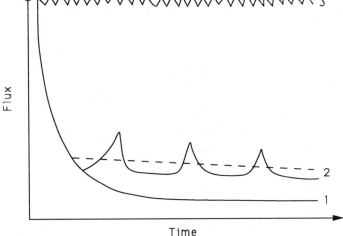

Figure 5.13. Effect of sequential backflushing on permeate flux (1: flux without backflush; 2: flux with a low backflush frequency; 3: flux with a high backflush frequency.

(50% increase or higher) the percentage loss in the permeate produced which is typically below 10% as illustrated in Figure 5.13.

5.8.6. Other System Design Considerations

There are many other important aspects that need to be considered in the filtration system design both on a pilot-scale and an industrial-scale. Some of these are the volumetric concentration factor, the presence of gas or vapor microbubbles, system drainage and air bleed.

The value of the volumetric concentration factor will depend on the system dead volume, the permeate volume and the ratio of bleed to permeate withdrawn. The presence of gas or vapor microbubbles can cause severe pore blockage. For instance, the formation of microbubbles inside the membrane layer and its support structure can lead to significant flux reduction even to the extent of total flux loss. The probability for the occurrence of such a situation increases as the membrane pore size increases. At higher pore diameters, air or gas can penetrate at lower pressures (e.g. 1–3 bar). Therefore, care must be taken to remove air or gas from the feed and circulation loop before filtration, to ensure that no air is drawn in or retained in the system.

Higher pressures on both sides of the membrane are used when desirable, as in filtrations involving gas-saturated liquids (e.g. beer, champagne, sparkling wine, etc.). In such applications, the permeate pressure is maintained above the saturation pressure of the gas. On the other hand, if solvents are used, precautions must to be taken to remove any dissolved air or gas as it will expand when the solvent flows across into the permeate side where the pressure is low. This type of fouling or pore clogging needs to be monitored not only during the normal filtration but also during backflushing.

When starting with a fresh dry membrane, it is also necessary to wet the membrane structure properly and completely. Wetting, especially with completely dry or new membranes, should be performed slowly and under as low a pressure as possible by using a clean liquid. The use of capillary adsorption is generally the most recommended technique to wet the membrane and fill the porosity.

It is also necessary to provide one or more bleed valves on the lowest points in the system both on the retentate and permeate sections. These will allow complete drainage without the creation of air pockets. Additionally, air bleeds are needed on the highest system points to allow for all entrapped air to escape prior to the start of the filtration operation.

When starting on a product, the shell side of the modules must be full of clean water (in aqueous systems) and the permeate valves closed. The retentate side is then filled with product and purged of air. When the desired velocity is achieved, the permeate valves can be opened slowly until the desired transmembrane pressure or filtrate flux is attained.

Mineral membranes could be sensitive to vibrations when operating horizontally. It is therefore necessary to ensure that the vibrations are not excessive (less than 50 kHz frequency) as to cause element breakage. Vertical modules installations are preferred which also improve air bleed and liquid drain when starting or stopping the filtration operation.

If the product is sensitive to shear stress, it is very important to choose the right pumps. For example, positive displacement pumps such as lobe and gear or centrifugal pumps with large closed wheel and low rotating speed are recommended. In multistage systems designed to produce a concentrated product, the last stage often includes positive displacement pumps and a larger membrane channel diameter to handle the higher retentate/product viscosity.

In many microfiltrations and also in some ultrafiltration applications, it is necessary to control the transmembrane pressure on each module to improve reliability and performance and to maintain high cleaning efficiency. Pumps transfer energy to the liquid which often increases the temperature of the filtration loop. Adding a heat exchanger allows one to control the operating temperature and to prevent modifications in the product characteristics. A heat exchanger also helps in the chemical cleaning cycle because, most often, the higher the temperature, the better the cleaning efficiency will be.

NOMENCLATURE

a, A	constants for a given liquid
dU/dr	shear rate
F_0	initial flux
F_α	filtrate flux when a fraction of pores (α) is blocked
n	number of monolayers of particles
P_1	tube-side (feed) inlet pressure
P_2	permeate outlet pressure
P_3	tube-side (retentate) outlet pressure
Q	total permeate rate, Equation (5.8), m^3/h
Q_1	permeate rate, m^3/h
Q_3	flow rate of retentate, Equation (5.7), m^3/h
t	duration of operation, h
V_0	system dead volume, m^3
V_c	volumetric concentration factor expressed as retentate concentration/feed concentration based on volume/flow considerations

Greek

α	fraction of the pores blocked
α'	minimum retention ratio, Equation (5.3)
α_1	initial permeate rate at $V_c = 1$, Equation (5.8)
α_2	experimentally determined parameter, Equation (5.8)
ΔP_B	pressure differential between the permeate-side and feed-side under the backflushing condition
ΔP_T	transmembrane pressure
τ	shear stress

REFERENCES

Bhave, R. R. and H. L. Fleming. 1988. Removal of oily contaminants in wastewater with microporous alumina membranes. *A.I.Ch.E. Symp. Ser.* **84**(261): 19–27.

Bhave, R. R., J. Gillot and P. K. T. Liu. 1989. High temperature gas separations for coal offgas cleanup using microporous ceramic membranes. Paper read at the 1st Intl. Conf. Inorganic Membranes, 3–7 July 1989, Montpellier.

Cartwright, P. S. 1985. Membrane separations technology for industrial effluent treatment— a review. *Desalination* **56**: 17–35.

Chen, S. C., J. T. Flynn, R. G. Cook and A. Cassiday. 1991. Removal of oil, grease and suspended solids from produced water using ceramic crossflow microfiltration. *SPE Production Engineering* (in press).

Cheryan, M. 1986. *Ultrafiltration Handbook.* pp. 151, 188 and 193. Technomic Publishing Co., Lancaster, PA.

Chong, R., P. Jelen and W. Wong. 1985. The effect of cleaning agents on a noncellulosic ultrafiltration membrane. *Sep. Sci. Technol.* **20**: 393–402.

Gach, G. J. 1986. Crossflow membrane filtration expands role in water treatment. *Power* **130**: 65–70.

Galaj, S., A. Wicker, J. P. Dumas, J. Gillot and D. Garcera. 1984. Crossflow microfiltration and backflushing on ceramic membranes. *Le Lait* **64**: 129–41.

Gillot, J. 1990. (personal communication).

Gillot, J., G. Brinkman and D. Garcera. 1984. New ceramic filter media for crossflow microfiltration and ultrafiltration. In *Filtra 84 Conference*, 2–4 October, Paris.

Gillot, J., M. Soria and D. Garcera. 1990. Recent developments in the Membralox® ceramic membranes. *Proc. 1st Intl. Conf. Inorganic Membranes*, 3–7 July, 379–81. Montpellier.

Gramms, L., D. Comstock and D. Hagen. 1984. Hydroperm lime softening—A case history. *Proc. 45th Intl. Water Conf.*, p. 41. Engineers Society of Western Pennsylvania, Pittsburgh, PA.

Hoogland, M. R., A. G. Fane and C. J. D. Fell. 1990. The effect of pH on the crossflow filtration of mineral slurries using ceramic membranes. *Proc. 1st Intl. Conf. Inorganic Membranes*, 3–7 July, 153–62. Montpellier.

Hsieh, H. P., R. R. Bhave and H. L. Fleming. 1988. Microporous alumina membranes. *J. Membrane Sci.* **39**: 221–41.

Lee, S., Y. Aurelle and H. Roques. 1984. Concentration polarization, membrane fouling and cleaning in ultrafiltration of soluble oil. *J. Membrane Sci.* **19**: 23–38.

Leenaars, A. F. M. and A. J. Burggraaf. 1985a. The preparation and characterization of alumina membranes with ultrafine pores. Part 3. The permeability for pure liquids. *J. Membrane Sci.* **24**: 245–66.

Leenaars, A. F. M. and A. J. Burggraaf. 1985b. The preparation and characterization of alumina membranes with ultrafine pores. Part 4. Ultrafiltration and hyperfiltration experiments. *J. Membrane Sci.* **24**: 261–70.

Malmberg, R. and S. Holm. 1988. Low bacteria skim milk by microfiltration. *North. Eur. Food Dairy J.* **1**: 75–77.

Matsumoto, Y., S. Nakao and S. Kimura. 1988. Crossflow filtration of solutions of polymers using ceramic microfiltration. *Intl. Chem. Eng.* **28**(4): 677–83.

Maubois, M. L. 1990. (personal communication).

Norton company. 1984. Ceraflo®: Testing tubular crossflow modules. Technical brochure.

Paulson, D. J., R. L. Wilson and D. D. Spatz. 1984. Crossflow membrane technology and its applications. *Food Technol.* **38**: 77–87.

Porter, M. C. 1986. Microfiltration. NATO ASI Ser. Vol. 181. Synthetic Membranes: Science, Engineering and Applications. ed. M. B. Chenoweth pp. 225–47

Sandblom, R. M. 1978. Filtering process. U.S. Patent 4,105,547.

Societe des Ceramiques Techniques (SCT)/Alcoa. 1987. *Membralox® Microfiltration and Ultrafiltration User's Manual*.

Tragardh, G. and P. Wahlgren. 1990. Removal of bacteria from beer using crossflow microfiltration. *Proc. 1st Intl. Conf. Inorganic Membranes*, 3–7 July, 291–95. Montpellier.

van Gassel, T. J. and S. Ripperger. 1985. Crossflow microfiltration in the process industry. *Desalination* **53**: 373–87.

Zydney, A. L. and C. K. Colton. 1986. A concentration polarization model for the filtrate flux in crossflow microfiltration in particulate suspensions. *Chem. Eng. Commun.* **47**: 1–21.

6. Gas Separations with Inorganic Membranes

R. J. R. UHLHORN and A. J. BURGGRAAF*

University of Twente, Faculty of Chemical Engineering, Enschede

6.1. INTRODUCTION

This chapter will deal with the application of inorganic membranes in gas separations. Because the separation mechanisms differ significantly, the inorganic membranes will be divided into two categories: porous inorganic membranes and nonporous ones. Examples of porous membranes are, e.g. Vycor glass, alumina membranes and carbon molecular sieve membranes and those for nonporous membranes are metal membranes and liquid-immobilized membranes (LIM), in which the liquid is a molten salt (Pez and Carlin 1986). Important transport and separation mechanisms in porous membranes are Knudsen diffusion (gas phase transport), surface diffusion (surface transport), multilayer diffusion and capillary condensation. When the pore size of the medium is of molecular dimensions, the transport mechanism is molecular sieving or micropore diffusion. In nonporous membranes the transport is by solution of the (gas) molecules in the membrane, followed by diffusion of the species through the membrane and finally dissolution.

The use of ceramic membranes in gas separations is not new. Since 1950, alumina membranes were used in the separation of UF_6 isotopes. However, the separation factor is very low in this case (theoretically 1.004!).

The interest in ceramic membranes grew, together with the interest in membrane separation processes, due to their specific properties. They are chemically stable, can withstand high temperatures and are noncompressible. These characteristics made them the only materials available, which could withstand the harsh environment in the isotope separation. On the other hand, the brittleness of most materials is a problem and so is the selectivity.

In order to make a good gas separation membrane, two demands should be met. First a high permeability is necessary and second the membrane should be selective. Porous membranes have quite high permeabilities [several times 10000 Barrer (1 Barrer = 1×10^{-10} cm³/cm²-s-cm Hg) for nitrogen (Vuren et al. 1987)], but relatively low selectivity. Nonporous mem-

* With R. R. Bhave.

155

branes show the opposite characteristics: high selectivities, but low permeabilities. In recent years improvements have been made in porous as well as nonporous systems, as will be shown in this chapter. The trend is towards high permeabilities and high selectivities, even at elevated temperatures, especially in microporous membranes (pores between 0.3 and 2 nm in diameter). Commercial inorganic gas separation membranes are still very scarce however. Metal membranes have been used to purify hydrogen, as early as 1965. Relatively new is the Al_2O_3 membrane of NGK used in dehydration applications (Abe 1986). This can be used to separate water from alcohols at temperatures up to 90°C. This way one can go beyond the azeotropic point and allow the production of higher purity alcohols.

This chapter will only deal with the possible gas transport mechanisms and their relevance for separation of gas mixtures. Beside the transport mechanisms, process parameters also have a marked influence on the separation efficiency. Effects like backdiffusion and concentration polarization are determined by the operating downstream and upstream pressure, the flow regime, etc. This can decrease the separation efficiency considerably. Since these effects are to some extent treated in literature (Hsieh, Bhave and Fleming 1988, Keizer et al. 1988), they will not be considered here, save for one example at the end of Section 6.2.1. It seemed more important to describe the possibilities of inorganic membranes for gas separation than to deal with optimization of the process. Therefore, this chapter will only describe the possibilities of the several transport mechanisms in inorganic membranes for selective gas separation with high permeability at variable temperature and pressure.

First, porous membranes will be discussed. Gases can be separated due to differences in their molecular masses (Knudsen diffusion), due to interaction (surface diffusion, multilayer diffusion and capillary condensation) and due to their size (molecular sieving). All these mechanisms and their possibilities will be discussed. For the sake of simplicity, theoretical aspects are not covered in detail, but examples of separations in literature will be given. The next section deals with nonporous membranes. Here the separation mechanism is solution–diffusion, e.g. solution and diffusion of hydrogen through a platinum membrane. This section is followed by an outline of some new developments and conclusions.

The literature cited is selected to give illustrative references to study the subject in more detail, without being complete.

6.2. POROUS MEMBRANES

6.2.1. Gas Separation by Knudsen Diffusion

In pressure-driven gas separation processes, several transport mechanisms can occur. These can be divided into gas phase transport and transport through interaction with the solid. The latter will be discussed in the following paragraphs.

Three mechanisms can occur in gas phase transport. The first is due to molecule–molecule collisions, taking place with conservation of the total amount of momentum. This is the so-called molecular diffusion. The second and the third are due to molecule–wall collisions. In molecule–wall collisions, the molecules lose momentum to the wall. If there is enough interaction between rebounded and adjacent molecules—which means the molecules must (statistically) collide at least as much with each other as with the wall— the momentum loss is progressively and smoothly transferred to the bulk of the gas. This is the so-called laminar flow or viscous flow regime. It differs from the molecular diffusion by the fact that in laminar flow there is no segregation of species and there is a loss of momentum. The third case is when there is no interaction between a rebounded and adjacent molecule—which means the molecules collide (statistically) much more with the wall than with each other. This regime is called Knudsen diffusion. In this regime, there are as many individual fluxes as there are species and these fluxes are independent of each other (in contrast to molecular diffusion).

Of these three mechanisms, i.e. molecular diffusion, laminar flow and Knudsen diffusion, only two are important in pressure-driven separations. These are laminar flow and Knudsen diffusion. These can be qualitatively understood as follows. If the molecules "see" each other much more than they see the pore wall (which means the mean free path of the molecules is much smaller than the mean pore radius), laminar flow and molecular diffusion are important. The laminar flow is much larger, however, and the molecular flow can be neglected (Present and de Bethune 1949). If the molecules see the pore wall much more than they see each other, only Knudsen diffusion will occur. Thus, the molecular diffusion can be neglected in all circumstances. From now on it will be assumed, that only laminar flow and Knudsen diffusion occur.

For pure gases, a good indication of which mechanism is dominant, Knudsen diffusion or laminar flow, is given by the Knudsen number:

$$Kn = \frac{\lambda}{r} \quad \text{with } \lambda = \frac{16\eta}{5\pi P_m} \sqrt{\frac{\pi R T}{2M}} \tag{6.1}$$

where λ is the mean free path, r the pore radius, η the gas viscosity, P_m the mean pressure, R the gas constant, T the temperature and M the molecular mass. Poiseuille flow through a porous medium, occurring for $Kn \ll 1$, is given by

$$F_{0,p} = \frac{\varepsilon \mu_p}{8RT} \frac{\bar{r}^2}{\eta L} P_m \tag{6.2}$$

where F_0 is the permeability, ε the porosity, μ_p a shape factor and L the thickness of the porous medium. Knudsen diffusion, occurring for $Kn \gg 1$, is given by

$$F_{0, Kn} = \tfrac{2}{3} \varepsilon \mu_k \, \bar{v} r / RTL \quad \text{and} \quad \bar{v} = \sqrt{\frac{8RT}{\pi M}} \tag{6.3}$$

where $\bar{v}$ is the mean molecular velocity, depending on the molecular mass. For $Kn \cong 1$ and for pure gases, Poiseuille flow and Knudsen diffusion are assumed to be additive. As can be seen from Equations (6.2) and (6.3), Poiseuille flow is nonseparative, while gas separation by Knudsen diffusion can be determined from the ratio of permeability of two gases A and B

$$F_0^A / F_0^B = \sqrt{M_B / M_A} \tag{6.4}$$

Thus, Knudsen diffusion can separate gases according to their molecular mass.

For gas mixtures, description of transport becomes more difficult. If only Knudsen diffusion occurs, the molecules do not see each other and for binary mixtures Equation (6.4) is valid. In the case of combined laminar flow and Knudsen diffusion, the gases interact. In literature several models were developed to describe combined flow and diffusion through porous media. Among the first descriptions was the one by Present and de Bethune (1949), who used a momentum balance approach to describe gas transport. This was also done by Rothfeld (1963), Gunn and King (1969) and Scott and Dullien (1962). Wakao, Otani and Smith (1965) developed a random pore model, which was consistent for compressed alumina powder pellets, but deviated markedly from their data on Vycor glass. Evans, Watson and Mason (1961, 1962) and Mason, Malinauskas and Evans (1967) developed the so-called dusty-gas model. In this model, the porous medium is depicted as huge gas molecules (hence dusty gas), fixed in time and space. This model, although fairly complicated, is commonly used to predict fluxes through porous media. A good introduction to gas phase transport through porous media is given by Cunningham and Williams (1980).

In literature, several authors have described gas phase transport through various membranes, for pure gases as well as for mixtures. Schultz and Werner (1982) measured permeabilities of several gases at room temperature through Vycor glass and found Knudsen diffusion behavior. Lee and Khang (1986) determined permeabilities of a wide variety of gases through a highly porous silicon-based membrane. Transport through these membranes is purely by Knudsen diffusion (Figure 6.1). Suzuki (1987) measured permeabilities of porous alumina membranes. At feed pressures of 5–75 cm Hg and 20°C he found Knudsen diffusion for hydrogen, helium, nitrogen and carbon

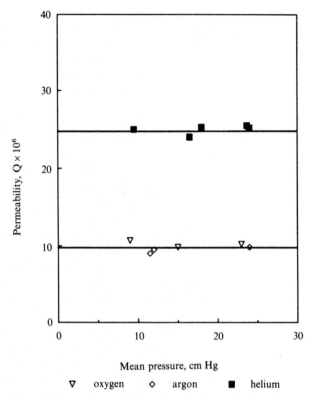

Figure 6.1. Permeability of several gases through a silica based hollow fiber membrane; data taken from Shindo et al. (1983).

dioxide. Itaya et al. (1984) measured gas transport characteristics of anodized alumina membranes and found Knudsen type behavior.

There are few studies in literature reporting pure gas permeabilities as well as separation factors of mixtures. Vuren et al. (1987) reported Knudsen diffusion behavior of pure gases for γ-alumina membranes with a mean pore radius of 1.2 nm. Separation experiments with a $1:1$ H_2/N_2 mixture showed, that the theoretical Knudsen separation factor [of 3.7, Equation 6.4)] for this mixture could be obtained (Keizer et al. 1988; see also Figure 6.2). In Figure 6.2, the effect of process parameters is also demonstrated. The separation factor is a function of the pressure ratio over the membrane, which is the ratio of the pressure on the permeate-side to that on the feed-side. For pressure ratios approaching unity, which means the pressure on both sides of the

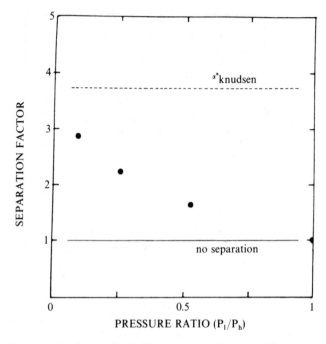

Figure 6.2. The separation factor of a H_2/N_2 mixture as a function of the pressure ratio at an average pressure near 100 kPa (Keizer et al. 1988).

membrane is almost equal, backdiffusion becomes important and the separation factor decreases. By decreasing the pressure ratio (increasing the difference between feed pressure and permeate pressure), the backdiffusion is also decreased and the separation factor approaches the ideal separation factor. In general it can be said, that the pressure ratio should be smaller than 0.1 in order to prevent backdiffusion.

Summarizing it can be stated that the separation by gas phase transport (Knudsen diffusion) has a limited selectivity, depending on the molecular masses of the gases. The theoretical separation factor is decreased by effects like concentration-polarization and backdiffusion. However, fluxes through the membrane are high and this separation mechanism can be applied in harsh chemical and thermal environments with currently available membranes (Uhlhorn 1990, Bhave, Gillot and Liu 1989).

6.2.2. Gas Separation by Surface Diffusion

Separation of gases by Knudsen diffusion is limited by the differences in molecular masses of the gases and is therefore only of practical importance

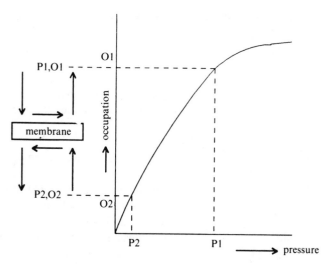

Figure 6.3. Difference in surface occupation (driving force for surface diffusion) as a function of the pressure at feed and permeate side of a membrane.

for the separation of light gases from heavier ones. For other mixtures, other separation mechanisms should be employed in order to obtain higher separation factors.

One of the alternatives is surface diffusion, well known in the fields of adsorbent technology and catalysis. Gas molecules can interact with the surface, adsorb on surface sites and be mobile on the surface. If a pressure gradient exists, a difference in surface occupation will occur (Figure 6.3). Therefore, a surface composition gradient is created and transport along the surface will occur, if the molecules are mobile. The gradient in surface diffusion is thus a *surface concentration* gradient. The concentration of adsorbed phase is a function of three parameters: the pressure, the temperature and the nature of the surface. However, there is a contradiction, since the more the molecules are adsorbed (the higher the heat of adsorption) the less the likelihood of their being mobile. This effect was clearly demonstrated by Uhlhorn, Keizer and Burggraaf (1989). It was shown that CO_2 exhibits surface diffusion on γ-alumina membranes. When the surface of the γ-alumina membrane was modified with magnesia, the amount of CO_2 adsorbed on the surface increased. The heat of adsorption also increased however and thus the mobility decreased. Due to this lower mobility, effectively less CO_2 was transported through the membrane, although more adsorption occurred.

There is another way to increase the contribution of the surface diffusion to total transport, in addition to altering the amount adsorbed and/or the surface mobility. This is through a pore size decrease. For cylindrical pores the following equation is commonly used to describe the relation between surface permeability and the structure of the porous medium:

$$F_{0s} \simeq \mu_s \varepsilon / r \tag{6.5}$$

For a single gas, if gas phase transport and surface transport are assumed to be additive, the following equation is valid, if it is assumed that $\mu_p = \mu_k = \mu_s$:

$$F_0 \cong F_{0,p} + F_{0,Kn} + F_{0,s} \cong \varepsilon \mu_k (r^2 + r + 1/r) \tag{6.6}$$

From Equation (6.6), it is immediately clear, that by decreasing the pore size, only the surface diffusion contribution is increased [third term on the right-hand side of Equation (6.6)], due to the increased surface area. It should be noted here that the additivity of surface permeability and gas phase permeability is still a matter of debate, since it is very difficult to determine the surface permeability alone. The experimentally measured surface permeability will always include a contribution due to gas phase permeability.

Surface diffusion has been extensively studied in literature. An overview of experimental data is given in Table 6.1. Okazaki, Tamon and Toei (1981), for example, measured the transport of propane through Vycor glass with a pore radius of 3.5 nm at 303 K and variable pressure (see Table 6.1). The corrected gas phase permeability was 0.69 m^3-m/m^2-h-bar, while the surface permeability was 0.55 m^3-m/m^2-h-bar, and so almost as large as the gas phase permeability (Table 6.1). It is clear from Table 6.1, that the effects of surface diffusion, especially at moderate temperatures, can be pronounced. At higher temperatures, adsorption decreases and it can be expected that surface diffusion will become less pronounced.

Several models were developed in literature to describe surface transport. These can be divided into three categories:

(1) The hydrodynamic model: In this model the adsorbed gas is considered as a liquid film, which can "glide" along the surface under the influence of a pressure gradient. Gilliland, Boddour and Russel (1958) used this model to calculate their fluxes.

(2) The hopping model: This model assumes that molecules can hop over the surface. The surface flux is calculated by the mean hopping distance and the velocity, with which the molecules leave their site. Weaver and Metzner (1966) developed a detailed model to calculate the mean hopping distance. Ponzi et al. (1977) developed a simpler way of estimating the mean hopping distance.

(3) The random walk model: This is the most frequently used model in literature. It is based on the two-dimensional form of Ficks law:

$$F_S = - A \rho_{app} D_s \mu_s \frac{dq}{dl} \tag{6.7}$$

where F_S is the surface flux, A the outer surface area, ρ_{app} the apparent density, D_S the surface diffusion coefficient and dq/dl the gradient in surface occupation. The surface diffusion coefficient has been the subject of many investigations. In most cases it is assumed that molecules jump from one site to another. This is an activated process and the energy of activation is a fraction of the heat of adsorption (Gilliland, Baddour and Perkinson 1974). This means that strongly adsorbed species are less mobile than weakly adsorbed species. If the activation energy is larger than 0.5 times the heat of adsorption, desorption occurs instead of surface diffusion (Okazaki et al. 1981). The model has been used by Ash, Barrer and Pope (1963), Carman and Raal (1950) and Tamon, Okazaki and Toei (1981).

Table 6.1. Surface Diffusion Data for Several Membrane/Gas Combinations; $P_{0,g}$ is the Corrected Gas Phase Permeability $(F_0 \times \sqrt{MT})$, $P_{0,s}$ the Corrected Surface Permeability

Author	Membrane	Radius (nm)	Gas	T(K)	Pressure (bar)	$P_{0,g}$ $\left(\dfrac{m^3 m}{m^2\text{-}h\text{-}bar}\right)$ $\times 10^3$	$P_{0,s}$ $\left(\dfrac{m^3 m}{m^2\text{-}h\text{-}bar}\right)$ 10^3
Horiguchi	Vycor	2.8	C_3H_6	273–323	0–1	0.63	0.47–2.0
			C_2H_6	273–323	0–1	0.63	0.2–0.9
	Graphon	10.6	He	273	0–1	11	4.4
			N_2	273–323	0–1	11	11–6
			C_3H_6	273–323	0–1	11	19–59
			C_2H_6	273–323	0–1	11	7.9–24
Stahl	Vycor	?	N_2	298	0–70	0.24	0.05
			CH_4	298	0–70	0.24	0.09
			CO_2	323	0–70	0.24	0.20
Gilliland	Vycor	2.3	NH_3	273–313	0–1	0.43	0.11
			CO_2	195–323	0–1	0.43	0.45–1.1
			SO_2	288–303	0–1	0.43	0.72–1.0
	Vycor	3.07	C_3H_6	273–323	?	0.58	0.27–1.1
			i-butane	273	?	0.58	0.6–9.4
Okazaki	Vycor	3.5	C_3H_6	303	0–1	0.69	0.55
Weaver	Vycor	2.0	i-butane	298	0–0.75	0.58	0.48–0.8
Ponzi	Carbon	8.75	freon	268–313	0–1	9.8	11.5–3.8
Rhim	VF-Millipore	20–30	C_3H_6	273	1–2	47	34
			O_2/N_2	273	1–2	47	0,9.0
			CH_4/CO_2	273	1–2	47	10,21
Tamon	Alumina	25	C_2H_4	283–303	0–1	11–31	5.2–0
			C_3H_6	283–303	0–1	11–40	17–4.6
			i-butane	283–303	0–1	11–51	25–3.6
Sladek	Porous Pt	2.5	H_2	333–348	0–0.003	0.24	0–0.72

All the experimental data in Table 6.1 refer to pure gases. Separation experiments, in which surface diffusion is the separation mechanism, are scarcely reported. Feng and Stewart (1973) and Feng, Kostrov and Stewart (1974) report multicomponent diffusion experiments for the system $He-N_2-CH_4$ in a γ-alumina pellet over a wide range of pressures (1–70 bar), temperatures (300–390 K) and composition gradients. A small contribution of surface diffusion (5% of total flow) to total transport could be detected, although it is not clear, which of the gases exhibits surface diffusion. The data could be fitted with the mass-flux model of Mason, Malinauskas and Evans (1967), extended to include surface diffusion.

Kameyama, Fukuda and Dokiya (1980, 1981) reported separation experiments of H_2-H_2S mixtures with porous glass (mean pore diameter 4.5 nm) and an alumina membrane (mean pore diameter 100 nm) up to temperatures of 800°C. They used the separation of H_2 from the mixture to shift the equilibrium in the decomposition reaction of H_2S. H_2S, however, diffused partly along the surface, especially at the lower temperature range, thereby decreasing the separation efficiency. No attempt was made to describe the transport. Shindo et al. (1983, 1984) reported data on the separation of $He-CO_2$ mixtures by means of a porous glass membrane over the temperature range 296–947 K. A theoretical analysis was made, studying the

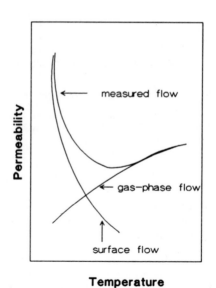

Temperature

Figure 6.4. Schematic view of total flow as sum of gas phase and surface flows; reproduced from Shindo et al. (1984).

diffusion flux as a function of temperature. Figure 6.4 shows in a general form the overall permeability as a function of pressure, reported by (Shindo et al. 1984). At low temperatures, surface flow is dominant, but with increase in temperature, the gas phase flow becomes dominant (Figure 6.4). Above 300°C surface diffusion was never high. It seems that at this temperature the surface flow had almost disappeared. This was also shown by Bhave, Gillot and Liu (1989), who measured carbon dioxide permeability on γ-alumina membranes (pore diameter 4.0 nm). Above 300°C the surface diffusion was only a few percent of total flow. Above 450°C surface flow was nonexistent.

In the cases reported previously, surface diffusion could have a pronounced effect and could increase the separation efficiency by a factor of 5. However, a sharp increase in the separation efficiency (a few orders of magnitude) is not to be expected, due to the conflict between adsorption and mobility. A high heat of adsorption will invariably lead to a low mobility. Another drawback of surface diffusion is that it becomes less effective at high temperatures. Therefore, it can be concluded that in order to employ surface diffusion as an effective separation mechanism, the pores should be very small (radius < 3 nm) and the temperature should be kept low ($T < 300$°C), due to the necessary physical adsorption of the gas (Uhlhorn 1990).

6.2.3. Gas Separation by Multilayer Diffusion and Capillary Condensation

An adsorbed phase can exhibit monolayer adsorption as well as multilayer adsorption. Surface flow in the presence of multilayer adsorption can be accounted for in the models described in the previous section. For example Okazaki and Tamon (1981) describe multilayer diffusion in their random hopping model.

Only a few studies have been published, showing capillary condensation. Although separation by capillary condensation is not new at all, but has been widely used in separation processes exploiting porous adsorbents, the dynamic behavior of flow of capillary condensate through porous media has received little attention. And it is this dynamic behavior that is important when capillary condensation is used as a separation mechanism.

The available experimental data and theoretical considerations are widely different (Gilliland and Baddour 1958, Eberly and Vohsberg 1965, Rhim and Hwang 1975, Tamon et al. 1981, Toei et al. 1983, Lee and Hwang 1986). It is difficult to compare them, since different methods of measuring permeability and different porous materials with incomplete characterization have been used. Lee and Hwang (1986) present a model, in which six different flow modes are considered (Figure 6.5). For each case, equations to calculate flow are given. This model is in good agreement with their experimental data for Freon-113 and water on Vycor glass with pores presumably around 4 nm

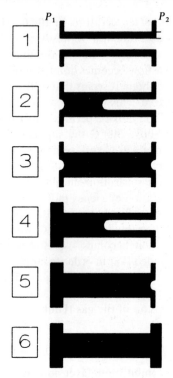

Figure 6.5. Six flow models for multilayer diffusion and capillary condensation; (1) multilayer diffusion, (2) capillary condensation at feed side, (3) entire pore filled with condensate, (4) bulk condensate at feed side, (5) bulk condensate at feed side and capillary condensate at permeate side and (6) total bulk condensate; data taken from Lee and Hwang (1986).

diameter. When the capillary condensation pressure in the pores is approached (at a relative pressure of 0.6), the permeability rapidly increases. After reaching a maximum, which is about 20–50 times higher than the gas phase permeability, the permeability drops due to blockage of the pores by condensate. It is one of the few models which accounts for the abrupt decrease in permeability at high relative pressures.

Very little data exist on the separation efficiency of multilayer diffusion and capillary condensation. Asaeda and Du (1986) used a thin modified alumina membrane to separate alcohol/water gaseous mixtures at high relative pressures (near 1). The azeotropic point could be bypassed for water/ethanol and water/isopropanol mixture by employing capillary condensation as a separation mechanism at a temperature of 70°C. By decreasing the pore size to the microporous range (pore diameter < 2 nm by plugging the pores with hydroxides), the separation factors were increased to above 60 (Asaeda and

Du 1986). This reflects the importance of small pores in order to apply effectively capillary condensation as a separation mechanism. Uhlhorn (1990) demonstrated the effect of multilayer diffusion of propylene through a modified γ-alumina membrane at 0°C. The separation factor for the N_2/C_3H_6 mixture was 27, where propylene is the preferentially permeating component, while the permeability increased to 7 times the Knudsen diffusion permeability. Although this mechanism appears to be very effective because of a high separation factor and a high permeability, it is limited by the obvious need for a condensable component. This in turn restricts the applicability range, due to limits set by temperature and pressure, needed for formation of multilayers or capillary condensation.

Finally it should be noted that inorganic membranes suitable for separation by multilayer diffusion and capillary condensation are also appropriate for performing pervaporation (a technique in which the feed is liquid and the permeate is gas) and distillation at reduced pressure (where gases with overlapping condensation regions are separated).

6.2.4. Gas Separation by Molecular Sieving

Molecular sieves are porous aluminosilicates (zeolites) or carbon solids that contain pores of molecular dimensions which can exhibit selectivity according to the size of the gas molecule. The most extensive study on carbon molecular sieve membranes is the one by Koresh and Soffer (1980, 1987). Bird and Trimm (1983) also described the performance of carbon molecular sieve membranes, but they were unable to prepare a continuous membrane. Koresh and Soffer (1980) prepared hollow-fiber carbon molecular sieves, with pores dimensions between 0.3 and 2.0 nm radius (see Chapter 2).

Table 6.2 shows some gas permeabilities and permselectivities for several gases through a membrane, activated at 950°C. From these data it is clear, that by simple thermochemical treatment, permeability and selectivity can be influenced. The product of permeability and selectivity is among the highest ever reported.

Recently Koresh and Soffer (1987) published data on the separation of a CH_4/H_2 mixture. They found a separation factor of about 30–50 for hydrogen, depending on temperature, and a permeability of about 2000 Barrer for hydrogen. The permeabilities of the gases in the mixture is equivalent to the permeabilities of the single gases. This means that the two gases behave independently. To account for this behavior, a qualitative pore constriction model has been developed. It is assumed that the H_2 and CH_4 molecules reside at different minimum energy positions prior to an activated jump through a pore constriction (Figure 6.6). The larger molecule will reside at a greater distance than a smaller molecule, due to the amorphous character of

Table 6.2. Permeability Data of a Carbon Molecular Sieve Membrane at 950°C for Several Modes of Activation (Pore Opening with Oxygen)

Activation	Gas	Permeability $(cm^3\text{-}cm/cm^2\text{-}s\text{-}cm\ Hg)$ $\times 10^8$	Selectivity
Nonactivated	He	10.8	
	CO_2	2.7	4.0
	O_2, N_2, SF_6	< 0.3	> 9
Activation	O_2	2.3	
(1st step)	N_2	< 0.3	> 8
Activation	He	52	
(2nd step)	O_2	17.1	3.0
	N_2	2.4	7.1
	SF_6	< 0.1	> 24
Degassing	He	44	
at 600°C	N_2	39	1.1
	O_2	91	2.3

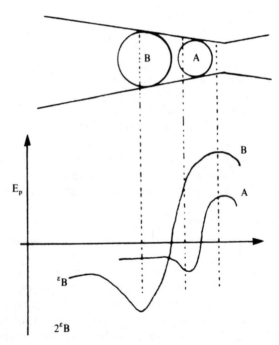

Figure 6.6. The potential energy (E_p) along the permeation path of two molecules of different sizes representing hydrogen and methane; reproduced from Koresh and Soffer (1987).

the membrane. Although this model can qualitatively explain the results, a more complete model to describe the permeation through molecular sieve membranes is needed.

6.3. DENSE MEMBRANES

In dense membranes, no pore space is available for diffusion. Transport in these membranes is achieved by the solution diffusion mechanism. Gases are to a certain extent soluble in the membrane matrix and dissolve. Due to a concentration gradient the dissolved species diffuses through the matrix. Due to differences in solubility and diffusivity of gases in the membrane, separation occurs. The selectivities of these separations can be very high, but the permeability is typically quite low, in comparison to that in porous membranes, primarily due to the low values of diffusion coefficients in the solid membrane phase.

Most separations with polymer membranes are based on the above described mechanism and extensive transport models have been developed in this field. Dense inorganic membranes however have also been in use for several decades. Since 1950, literature is available on hydrogen diffusion through thin platinum foils. The selectivity is very high (> 1000), but the permeability is low (about 2000 Barrer). Early in the 1960s commercially operating plants produced pure hydrogen by using palladium foils (Price 1965). Extensive studies describing the mechanism of hydrogen dissolution and diffusion have been performed (de Rosset 1960, Barner 1951, Smithells 1937). The diffusion rate is dependent on pressure according to

$$D = kP^a \qquad \text{with } 0.5 < a < 1.0 \tag{6.8}$$

where k is a constant. A real breakthrough with this type of membrane separation has yet to be achieved, mainly due to high fabrication and energy costs, and low stability and volume flow (Rödicker 1974).

Diffusion of oxygen through silver membranes has also been reported (Yunoshin 1967). It was however necessary to incorporate additives to stabilize the silver foils, because pure silver is not stable in oxidizing environments.

An alternative use of metal membranes has been demonstrated in literature Gryaznov et al. (1980), Gryaznov (1986). Alloys, based on palladium were used (Gryaznov and Smirnov 1977), as membrane catalysts, which are selectively permeable to hydrogen. These membrane catalysts are active and selective with respect to reactions involving the splitting-off or addition of hydrogen. It could be shown that yields and selectivities of catalytic hydrogenation products were improved as a result of the controlled introduction of hydrogen through the metal membrane (Gryaznov 1986). In dehydrogenation reactions the product yield is increased through the preferential removal of hydrogen (see also Chapter 7).

6.4. NEW DEVELOPMENTS

In this last section some recent developments are mentioned in relation to gas separations with inorganic membranes. In porous membranes, the trend is towards smaller pores in order to obtain better selectivities. Lee and Khang (1987) made microporous, hollow silicon-based fibers. The selectivity for H_2 over N_2 was 5 at room temperature and low pressures, with H_2 permeability being 2.6×10^4 Barrer. Hammel et al. 1987 also produced silica-rich fibers with mean pore diameter 0.5–3.0 nm (see Chapter 2). The selectivity for helium over methane was excellent (500–1000), but permeabilities were low (of the order of 1–10 Barrer).

Uhlhorn et al. (1989) have studied a composite alumina membrane, of which the top layer is microporous silica. This top layer is only 30–60 nm thick and thermally stable at temperatures up to 800°C, provided no water vapor is present (unlike, e.g. the alumina membrane of Asaeda and Du 1986). The pore diameter is of molecular dimensions, typically 0.5–1.0 nm, and can be varied by the synthesis. Selectivities for several gases at different temperatures and a moderate pressure were determined. These appear to be very good. For CO_2/CH_4 mixture the selectivity was 60 at 100°C, where CO_2 is the preferentially permeating component, and for H_2/C_3H_6 mixture it was 260 at 300°C, where hydrogen is the preferentially permeating component. Permeabilities were also good (several thousands of Barrer, e.g. for hydrogen, which is only one order of magnitude smaller than the Knudsen diffusion permeability). Silica membranes containing micropores were also prepared by Gavalas, Megiris and Nam (1989), yielding a membrane with high hydrogen selectivity (separation factor for a H_2/N_2 mixture was 2000), but very low permeabilities. Okubo and Inoue (1989) synthesized silica-modified glass membranes and showed the occurrence of micropores, but have not reported their gas separation performance.

Also very interesting is the development of a zeolite membrane by Suzuki (1987), consisting of a support and an ultrathin film of a cage shaped zeolite (see also Chapter 2). Pore diameters ranged from 3–12 Å and can be varied by synthesis. There are numerous kinds of possibilities for this kind of membrane, not only for separation, but also for catalytic applications. It was shown, for example that a $1:1:1$ $CH_4:C_2H_6:C_3H_8$ could be separated at room temperature and 15 bar pressure to give a 63.8 wt.% CH_4, 26 wt.% C_2H_6 and 0.5 wt.% C_3H_8 mixture in one step. Permeability data however are not given.

Some developments in dense membranes have also been reported. Uemiya et al. (1988), for example, decreased the thickness of a palladium membrane by applying it on Vycor glass. This increased the permeability by a factor of 3 while letting only hydrogen permeate. The coating of porous substrates with Pd has been reported frequently in recent years (e.g. Konno et al. 1988), mostly for use in catalytic membrane reactors. Another good example of a new dense membrane is the LIM of Pez and Carlin (1986), in which the liquid is a molten salt (see Chapter 2). Transport occurs through reversible reactions. At the feed-side, gas uptake occurs by reaction and due to a concentration gradient, the product of this reaction diffuses to the permeate-side, where the reversible reaction regenerates the gas. With this principle, N_2–NH_3 mixtures could be separated at 300°C with a selectivity of 30–245 for ammonia over N_2, and a NH_3 permeability of 10^4 Barrer. O_2–N_2 separation at 430°C gave a selectivity of 172 for O_2 over N_2 and a O_2 permeability of about 2000 Barrer.

Finally the synthesis of inorganic-polymer composite membranes should be mentioned. Several attempts have been made to combine the high permeability of inorganic membranes with the good selectivity of polymer membranes. Furneaux and Davidson (1987) coated a anodized alumina with polymer films. The permeability increased by a factor of 100, as compared to that in the polymer fiber, but the selectivities were low ($H_2/O_2 = 4$). Ansorge (1985) made a supported polymer film and coated this film with a thin silica layer. Surprisingly, the silica layer was found to be selective for the separation mixture He–CH_4 with a separation factor of 5 towards CH_4. The function of the polymer film is only to increase the permeability. No further data are given.

In summary, it may be stated that thin, microporous membranes offer by far the largest selectivities and reasonable fluxes in a wide range of temperatures, but the material processing is not yet fully understood and controlled (Uhlhorn 1990).

6.5. CONCLUSIONS

It may be useful to summarize the main highlights and future prospects in the development of inorganic gas separation membranes.

First of all it is clear, that inorganic membranes with good selectivity as well as good permeability are yet available, but progress has been made in their development. The most promising development is the development of microporous membranes with pore sizes between 0.5 and 2.0 nm. Mechanisms that will give high permeabilities and selectivities in this kind of membranes, are multilayer diffusion/capillary condensation, molecular sieving or activated surface transport, where molecules are transported by activated jumps through pore constrictions (e.g. Koresh and Soffer 1987). The two last mentioned mechanism have the advantage that they are also effective at high temperatures (> 600 K). The development of the carbon molecular sieve membranes, the zeolite membrane and the various microporous (composite) membranes, based on silica, appear to suggest that efficient gas separation over a wide range of temperature and pressure may be possible in the near future.

Dense inorganic membranes can be interesting, if the thickness can be reduced sufficiently. Metal membranes however will continue to be very sensitive to their environment. LIMs, also working at high temperatures, seem a promising substitute. Finally, integration of polymer and inorganic membranes can give high selectivities and high permeabilities, but chemical and thermal stability of the composite membrane will probably be lower.

From a commercial point of view, inorganic membranes will become increasingly interesting. In 1989, the inorganic membrane market was estimated to be \$31 million, of which \$28 million was in ultra- and microfiltration (Egan 1989). By 1999 the total is projected to reach \$432 million, of which \$80 million will be in gas separation, primarily in refineries (Crull 1989).

From a material point of view, it can be expected that, in the near future, gas separation membranes with good selectivities and high permeabilities will become available. As interest in inorganic membranes is growing, the understanding of transport in relation to structure will increase, leading to a better control of separation properties. As the thermal and chemical stability will continue to be a strong advantage of inorganic membranes, their application will mainly be in harsh chemical environments or in high-temperature processes. It is however not only the membrane performance that will decide if gas separation by inorganic membranes can be industrially applied. Membrane modules, resistant to extreme conditions, must be constructed. Already now, one should become aware of the necessity of these developments and efforts should be put into membrane module construction, to keep pace with the improvements in membrane performance.

NOMENCLATURE

A	outer membrane surface, m^2
D_s	surface diffusion coefficient, $m^2\text{-s}^{-1}$
F_0	permeability, $mol/m^2\text{-s-Pa}$
J	flux, $mol\text{-m}^{-2}\text{-s}^{-1}$
Kn	Knudsen number,
L, l	membrane thickness, m
M	molar mass, $kg\text{-mol}^{-1}$
P_0	corrected permeability, $m^3\text{-m-m}^{-2}\text{-h}^{-1}\text{-bar}^{-1}$

P pressure, N-m^{-2}
q surface occupation, mol-kg^{-1}
r pore radius, m
R gas constant, 8.314 J-mol^{-1}-K^{-1}
T temperature, K
$\bar{v}$ mean molecular speed, m-s^{-1}

Greek

ε porosity
η viscosity, N-s-m^{-2}
λ mean free path, m
μ shape factor
ρ_{app} apparent density, kg-m^{-3}

Indices

Kn Knudsen regime
P Poiseuille flow regime
g gas phase
s surface phase

REFERENCES

Abe, F. 1986. A separation membrane and process for manufacturing the same. European Patent Appl. 0,195,549.
Ansorge, W. 1985. Membrane for separation of gases from gas mixture and method of its preparation. German Patent 3,421,833A1.
Asaeda, M. and L. D. Du. 1986. Separation of alcohol/water gaseous mixtures by a thin ceramic membrane. *J. Chem. Eng. Japan* 19(1): 72–77.
Asaeda, M. and L. D. Du. 1986. Separation of alcohol/water gaseous mixture by an improved ceramic membrane. *J. Chem. Eng. Japan* 19(1): 84–85.
Ash, R., R. M. Barrer and C. G. Pope. 1963. Flow of adsorbable gases and vapors in a microporous medium. *Proc. Roy. Soc.* A271: 1–18.
Barner, R. M. 1951. *Diffusion in and Through Solids.* University Press, New York.
Bhave, R. R., J. Gillot and P. K. T. Liu. 1989. High temperature gas separations for coal offgas cleanup with microporous ceramic membranes. Paper 124f read at AIChE Annual Meeting, 5–10 November 1989, San Francisco.
Bird, A. J. and D. L. Trimm. 1983. Carbon molecular sieves used in gas separation membranes. *Carbon* 21(3): 177–80.
Carman, P. C. and F. A. Raal. 1950. Capillary condensation in physical adsorption. *Proc. Roy. Soc.* A203: 165–178.
Crull, A. 1989. *Inorganic Membranes: Markets, Technologies, Players.* Business Communications Company., Inc. Norwalk, CT.
Cunningham, R. E. and R. J. J. Williams. 1980. *Diffusion in Gases and Porous Media.* Plenum Press, New York.

de Rosset, A. J. 1960. Diffusion of H_2 through Pd membranes. *Ind. Eng. Chem.* 52(6): 525–28.

Eberley, P and D. Vohsberg. 1965. Diffusion of C_6H_6 and inert gases through porous media at elevated temperatures and pressures. *Trans. Faraday Soc.* 61: 2724–35.

Egan, B. Z. 1989. Using Inorganic Membranes to Separate Gases: R&D Status Review. *Internal report ORNL/TM-11345*, Oak Ridge National Laboratory.

Evans, R. B., G. M. Watson and E. A. Mason. 1961. Gaseous diffusion in porous media at uniform pressure. *J. Chem. Phys.* 35: 2076–83.

Evans, R. B., G. M. Watson and E. A. Mason. 1962. Gaseous diffusion in porous media: (II) effect of pressure gradients. *J. Chem. Phys.* 36: 1894–902.

Feng, C. and W. E. Stewart, 1973. Practical models for isothermal diffusion and flow of gases in porous solids. *Ind. Eng. Chem. Fundam.* 12(2): 143–47.

Feng, C., V. V. Kostrov and W. E. Stewart. 1974. Multicomponent diffusion of gases in porous solids. Models and experiments. *Ind. Eng. Chem. Fundam.* 13(1): 5–9.

Furneaux, R. C. and A. P. Davidson. 1987. Composite membranes. European Patent. Appl. 242,209.

Gavalas, G. R., C. Megiris and S. W. Nam. 1989. A novel composite inorganic membrane for combined catalytic reaction and product separation. *Chem. Eng. Sci.* 44(9): 1829–35.

Gilliland, E., R. F. Baddour and J. L. Russel. 1958. Rates of flow through microporous solids. *A.I.Ch.E. J.* 4: 90–96.

Gilliland, E., R. F. Baddour and G. P. Perkinson. 1974. Diffusion on surfaces. Effect of concentration on the diffusivity of physically adsorbed gases. *Ind. Eng. Fundam.* 13: 95–100.

Gryaznov, V. M. 1986. Surface catalytic properties and hydrogen diffusion in palladium alloy membranes. *Z. Phys. Chem. Neue Folge* 147: 761–70.

Gryaznov, V. M. and Smirnov. 1977. Selective hydrogenation on membrane catalysts. *Kinetics and Catalysis* 3: 485–87.

Gryaznov, V. M., M. M. Ermilova, L. D. Gogua and S. I. Zavodchenko. 1980. Gas permeability of composite membrane catalysts. *Bulletin Academy Sciences USSR, Division of chemical sciences* 29(4): 529–32.

Gunn, R. D. and C. J. King. 1969. Mass transport in porous materials under combined gradients of composition and pressure. *A.I.Ch.E. J.* 15: 507–14.

Hammel, J. J., W. P. Marshall, W. J. Robertson and H. W. Barch. 1987. Porous inorganic siliceous-containing gas enriching material and process of manufacture and use. European Patent. Appl. 248,391.

Hammel, J. J., W. P. Marshall, W. J. Robertson and H. W. Barch. 1987. Porous inorganic siliceous-containing gas enriching material and process of manufacture and use. European Patent. Appl. 248,392.

Hsieh, H. P., R. R. Bhave and H. L. Fleming. 1988. Microporous alumina membranes. *J. Membrane Sci.* 39: 221–43.

Itaya, K., S. Sugawara, K. Arai and S. Saito. 1984. Properties of porous anodic aluminum oxide films as membranes. *J. Chem. Eng. Japan* 17(5): 514–20.

Kameyama, T., K. Fukuda and M. Dokiya. 1980. Production of hydrogen from hydrogen sulfide by means of selective diffusion membranes. In *Hydrogen Energy Progress, Proc. 3rd World Hydrogen Energy Conference*, eds. T. N. Veziroglu, K. Fueki and T. Ohta, pp. 569–79, Tokyo, Japan.

Kameyama, T., K. Fukuda and M. Dokiya. 1981. Possibility for effective production of hydrogen from hydrogen sulfide by means of a porous Vycor glass membrane. *Ind. Eng. Chem. Fundam.* 20: 97–99.

Keizer, K., R. J. R. Uhlhorn, R. J. van Vuren and A. J. Burggraaf. 1988. Gas separation mechanisms in microporous modified γ-Al_2O_3 membranes. *J. Membr. Sci.* 39: 285–300.

Konno, M., M. Shindo, S. Sugawara and S. Saito. 1988. A composite palladium and porous aluminum oxide membrane for hydrogen gas separation. *J. Membr. Sci.* 37: 193–97.

Koresh, J. E. and A. Soffer. 1980. Study of molecular sieve carbons. Part I. Pore structure, gradual pore opening and mechanism of molecular sieving. *J. Chem. Soc. Faraday. Trans.* 76(2): 2457–71.

Koresh, J. E. and A. Soffer. 1980. Study of molecular sieve carbons. Part II. Estimation of cross-sectional diameters of non-spherical molecules. *J. Chem. Soc. Faraday. Trans.* 76(2): 2472–85.

Koresh, J. E. and A. Soffer. 1980. Molecular sieving range of pore diameters of adsorbents. *J. Chem. Soc. Faraday. Trans.* 76(2): 2507–09.

Koresh, J. E. and A. Soffer. 1983. Molecular sieve carbon permselective membrane. Part I. Presentation of a new device for gas mixture separation. *Sep. Sci. Technol* 18(8): 723–34.

Koresh, J. E. and A. Soffer. 1987. The carbon molecular sieve membranes. General properties and the permeability of CH_4/H_2 mixture. *Sep. Sci. Technol.* 22(2): 973–82.

Lee, K. H. and S. T. Hwang. 1986. The transport of condensible vapors through a microporous Vycor glass membrane. *J. Colloid, Interf. Sci.* 110(2): 544–55.

Lee, K. H. and S. J. Khang. 1986. A new silicon-based material formed by pyrolysis of silicon rubber and its properties as a membrane. *Chem. Eng. Commun.* 44: 121–32.

Lee, K. H. and S. J. Khang. 1987. High temperature membrane. U.S. Patent 4,640,901.

Mason, E. A., A. P. Malinauskas and R. B. Evans. 1967. Flow and diffusion of gases in porous media. *J. Chem. Phys.* 36: 3199–216.

Okazaki, M., H. Tamon and R. Toei. 1981. Interpretation of surface flow phenomenon of adsorbed gases by hopping model. *A.I.Ch.E. J.* 27(2): 271–77.

Okubo, T. and H. Inoue. 1989. Introduction of specific gas selectivity to porous glass membranes by treatment with tetra ethoxy silane. *J. Membrane Sci.* 42: 109–117.

Pez, G. and M. Carlin. 1986. Method for gas separation. European Patent Appl. 0,194,483.

Ponzi, M., J. Papa, J. B. Rivarola and G. Zgrablich. 1977. On the surface diffusion of adsorbable gases through microporous media. *A.I.Ch.E. J.* 23: 347–52.

Present, R. D. and A. J. de Bethune. 1949. Separation of a gas mixture flowing through a long tube at low pressure. *Phys. Rev.* 75: 1050–57.

Price, F. C. 1965. Palladium diffusion yields high-volume hydrogen. *Chem. Eng.* 72(5): 36–38.

Rhim, H. and S. T. Hwang. 1975. Transport of capillary condensate. *J. Colloid. Interf. Sci.* 52: 174–81.

Rothfeld, L. D. 1963. Gaseous counterdiffusion in catalyst pellets. *A.I.Ch.E. J.* 9: 19–24.

Rödicker, H. 1974. Separating gas mixtures using metal coated membranes. East German Patent 107,859.

Schultz, G. and U. Werner. 1982. Membrane for gas separation. *Vt. Verfahrenstechnik* 16(4): 244–46.

Scott, D. S. and F. A. L. Dullien. 1962. Diffusion of ideal gases in capillaries and porous solids. *A.I.Ch.E. J.* 8: 113–17.

Shindo, Y., T. Hakuta, H. Yoshitome and H. Inoue. 1983. Gas diffusion in microporous media in Knudsen regime. *J. Chem. Eng. Japan* 16(2): 120–26.

Shindo, Y., T. Hakuta, H. Yoshitome and H. Inoue. 1984. Separation of gases by means of a porous glass membrane at high temperatures. *J. Chem. Eng. Japan* 17(6): 650–52.

Smithells, C. J. 1937. *Gases and Metals.* John Wiley & Sons, New York.

Suzuki, F., K. Onozato and Y. Kurokawa. 1987. Gas permeability of a porous alumina membrane prepared by the sol–gel process. *J. Non-Cryst. Solids* 94: 160–62.

Suzuki, H. 1987. Composite membrane having a surface layer of an ultrathin film of cage-shaped zeolite and processes for production thereof. U.S. Patent 4,699,892.

Tamon, H, S. Kyotani, H. Wada, M. Okazaki, and R. Toei. 1981. Surface flow phenomenon of adsorbed gases in porous media. *J. Chem. Eng. Japan* 14(2): 136–41.

Tamon, H., M. Okazaki and R. Toei. 1981. Flow mechanism of adsorbate through porous media in presence of capillary condensation. *A.I.Ch.E. J.* 27(2): 271–76.

Toei, R., H. Imakoma, H. Tamon and M. Okazaki. 1983. Water transfer coefficient in adsorptive porous body. *J. Chem Eng. Japan* 16(5): 364–69.

Uemiya, S., Y. Kude, K. Sugino, N. Sato, T. Matsuda and E. Kirkuchi. 1988. A palladium/ porous-glass composite membrane for hydrogen separation. *Chem. Lett.* 10: 1687–90.

Uhlhorn, R. J. R. 1990. Ceramic Membranes for Gas Separation: Synthesis and Gas Transport Properties, Ph.D. thesis, University of Twente, Enschede.

Uhlhorn, R. J. R., K. Keizer and A. J. Burggraaf. 1989. Gas and surface diffusion in (modified) γ-alumina systems. *J. Membrane Sci.* 46: 225–46.

Uhlhorn, R. J. R., M. H. B. J. Huis in't Veld, K. Keizer and A. J. Burggraaf. 1989. High permselectivities of microporous silica modified γ-alumina membranes *J. Mater. Sci. Lett.* 8: 1135–38.

van Vuren, R. J., B. Bonekamp, R. J. R. Uhlhorn, K. Keizer, H. J. Veringa and A. J. Burggraaf. 1987. Formation of ceramic alumina membranes for gas separation. In *Materials Science Monographs 38c (High Tech Ceramics)*, ed. P. Vincenzini, pp. 2235–46., Elsevier, Amsterdam.

Wakao, N., J. Otani and J. M. Smith. 1965. Significance of pressure gradients in porous materials Part I: diffusion and flow in fine capillaries. *A.I.Ch.E. J.* 11: 435–39.

Wakao, N., J. Otani and J. M. Smith. 1965. Part II: diffusion and flow in porous catalyst. *A.I.Ch.E. J.* 11: 439–45.

Weaver, J. A. and A. B. Metzner. 1966. The surface transport of adsorbed molecules. *A.I.Ch.E. J.* 12: 655–61.

Yunoshin, I. 1967. A silver membrane for oxygen diffusion cell. West German Patent 1,244,738.

7. Inorganic Membrane Reactors to Enhance the Productivity of Chemical Processes

V. T. ZASPALIS and A. J. BURGGRAAF*

University of Twente, Faculty of Chemical Engineering, Enschede

7.1. INORGANIC MEMBRANES FOR HIGH-TEMPERATURE APPLICATIONS

It was some fifteen to twenty years ago that the first papers appeared in the literature, pointing out that the application of membranes in reaction engineering would lead to the production of new chemical processes (Michaels 1968, Gryaznov 1970, Raymont 1975, Hwang and Kammermeyer 1975). The main idea was to prevent the reaction mixture from attaining the equilibrium composition by incorporating a membrane within the reactor, the membrane being used for the continuous and selective removal of products from the reaction zone. Inorganic membranes (ceramic or metallic, porous or dense), with their inherent thermal, structural and chemical stability, were in particular attractive for many high-temperature separations, for heterogeneous or homogeneous reactions or for reactions in aggressive and corrosive environments.

Since then a lot of work has been done on membrane reactor technology, new concepts and ideas have been investigated and advantageous properties have been experimentally realized, so that inorganic membrane reactor technology is considered in our days as a generic alternative to conventional reactor design in high-temperature chemical engineering applications. Several reviews have been published (Hsieh 1988, 1989, Lee 1987, Armor 1989, Ilias and Govind 1989, Dellefield 1988).

The broadening of the application field and the subsequent increased use of membranes has led to an expanded base of knowledge in their fabrication technology (i.e. Leenaars, Burggraaf and Keizer 1987, Suzuki 1987, Keizer and Burggraaf 1988, Burggraaf and Keizer 1990, Anderson, Gieselmann and Xu 1988, Furneaux and Thornton 1988, Guizard et al. 1989). However, the production costs of inorganic membranes and the fabrication of industrial-scale modules are still critical. Brittleness and surface integrity, reproducibility, high-temperature sealing and mass production are also critical production factors. Though the operation of membrane reactor processes

* With R. R. Bhave.

seems to be economically favorable (Michaels 1968, Raymont 1975, Lee 1987, Hurly 1987, Haggin 1988), any attempt to penetrate in the field of reaction engineering, where large-scale nonmembrane solutions have already been subjected to optimization and continuous improvement for many decades, would require the fabrication of larger membrane modules at lower production costs.

According to the most general definition, membranes are thin barriers that allow preferential passage of certain species across their structure. The way in which this selective transport occurs may vary depending on the structural and chemical characteristics of the membrane and on the nature of the specific interactions between the membrane surface and the phase through which transport across the membrane barrier occurs (Uhlhorn et al. 1989a).

Inorganic membranes can be conveniently divided into two main categories; dense and porous membranes. Both types have been extensively used in membrane reactor studies. Commonly used materials for dense membrane preparation are palladium or palladium-based alloys, platinum, nickel, silver or even ceramic materials such as stabilized zirconias (Hsieh 1989). Certain gases (e.g. hydrogen through palladium membranes or oxygen through silver or oxide membranes) can be transported via diffusion across their structure, e.g. solution/diffusion mechanism (Lakshminarayanaiah 1969). Porous membranes are currently made from glass, alumina, zirconia, titania, silver and stainless steel (Hsieh 1989) with pores which may vary from 40 Å to 10 μm. Recently, porous ceramic membranes incorporating silica have been reported (Uhlhorn et al. 1989b). Gases can be transported through the membrane pores via bulk flow, molecular or Knudsen diffusion, surface diffusion, multilayer adsorption and capillary condensation or microporous activated diffusion (Uhlhorn et al. 1989b, Gavalas, Megiris and Nam 1989). Porous membranes have been put to commercial use rather than the dense ones with the exception of the palladium membranes (Gryaznov and Karavanov 1979).

Inorganic membranes employed in reaction/transport studies were either in tubular form (a single membrane tube incorporating an inner tube side and an outer shell side in double pipe configuration or as multichannel monolith) or plate-shaped disks as shown in Figure 7.1 (Shinji et al. 1982, Zaspalis et al. 1990, Cussler 1988). For increased mechanical resistance the thin porous (usually mesoporous) membrane layers are usually supported on top of macroporous supports (pores 1–15 μm), very often via an intermediate porous layer, with pore size 100–1500 nm, (Keizer and Burggraaf 1988).

Dense metallic membranes have the advantage of very high selectivities since only certain species can be dissolved in their structural lattice. However, the permeabilities are lower by a factor of 100 than those of porous membranes (Ilias and Govind 1989, van Vuren et al. 1987, Itoh 1987, Suzuki, Onozato and Kurokawa 1987). For example, the permeability of

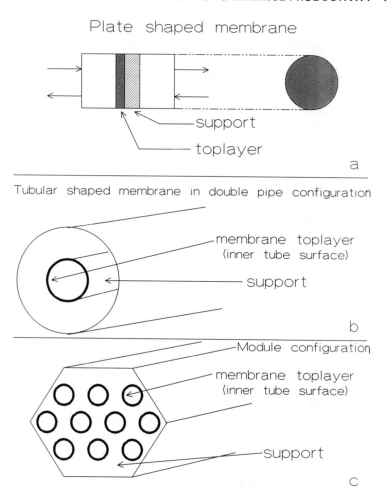

Figure 7.1. Typical membrane reactor configurations (a) reactor with plate-shaped membranes, (b) tubular-shaped membrane in double pipe configuration and (c) multichannel monolith.

hydrogen through a γ-Al_2O_3 membrane with an average pore size of 3 nm and a thickness of 4 μm at 20°C is 15×10^{-6} mol-m^{-2}-s^{-1}-Pa^{-1}, while the permeability of a 20 μm thick palladium foil at the same temperature is 6×10^{-8} mol-m^{-2}-s^{-1}-Pa^{-1}. For a better comparison, it has to be noted that the permeabilities through palladium membranes vary with the square root of the pressure difference and not linearly as is the case with porous membranes. Metallic membranes very often suffer from structural degradation particularly after repeated adsorption/desorption cycles at higher temperatures (above 200°C). Steam atmospheres and extreme pH conditions (pH lower than 2 or higher than 12) might cause structural instabilities also in dense or porous ceramic oxide membranes (Gavalas, Megiris and Nam 1989, Nam and Gavalas 1989, Wu et al. 1990). Possibilities for achieving dense membranes with high permeabilities and high selectivities seem to exist in the field of very thin film (nano-scale) preparation by chemical vapor deposition techniques (Lin, de Vries and Burggraaf 1989, Burggraaf and Keizer 1990), or in the synthesis of porous matrices which contain an immobilized-liquid phase, i.e. molten salt membranes (Burggraaf and Keizer 1990).

7.2. GAS (OR VAPOR) PHASE REACTIONS: THE CONCEPT OF THE MEMBRANE REACTOR

The use of membranes in reaction processes can serve two different purposes (1) to enhance the productivity of a chemical reactor, i.e. by shifting the chemical equilibrium situation. (2) to influence the path of a chemical reaction, i.e. by affecting the reaction selectivity.

The combination of inorganic membranes and reactors can be done in various ways as shown in Figure 7.2:

1. The membrane is a unit of the process separated from the reactor. It maintains only the separation function (passive membrane) and there is almost no interaction between reaction and separation. In fact, we have here two different processes connected in series. Stream purification from catalyst poisoning substances or feed enrichment of a recycle stream belong to the possibilities of this configuration.

2. The membrane forms the wall of the reactor where a homogeneous or a heterogeneous reaction takes place (Figure 7.2a). This operation mode provides a stronger interaction between reaction and separation, allowing the two processes to be integrated in one unit operation. The reaction is directly influenced by the membrane with the continuous and selective removal of product from the reaction zone or with the controlled supply of reactant to the reaction zone (passive membrane). This configuration has been extensively investigated in combination with catalytic reactions which are equilibrium limited to low conversions (e.g. Kameyama et al. 1979, 1981a, b, Shinji, Misono and Yoneda 1982, Itoh et al. 1984, Nagamoto and Inoue 1985, 1986, Itoh et al. 1988, Wu et al. 1990).

3. The membrane is inherently catalytic (Figure 7.2b) or modified with catalytically active species distributed in or at the entrance of the membrane pores as individual particles or as a layer (Figure 7.2c). The catalytic activity is adjusted to the membrane (catalytically active membrane). In this way the strongest interaction between membrane transport properties and catalytic activity can be achieved.

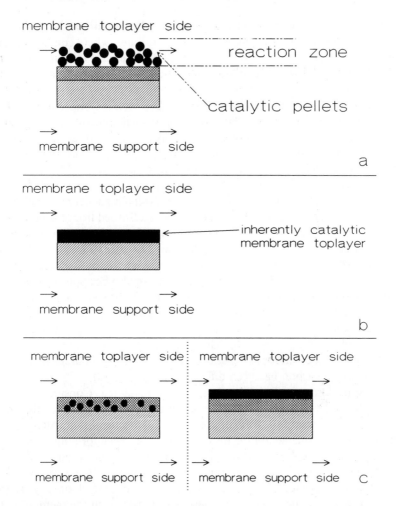

Figure 7.2. Different membrane/catalyst combinations (a) the catalyst is packed next to the membrane, (b) the membrane is inherently catalytic and (c) the membrane is modified with catalytically active components.

There are certainly quite significant advantages that membrane reactor processes provide as compared to conventional reaction processes. The reactor can be divided by the membrane into two individual compartments. The bulk phases of the various components or process streams are separated. This is of importance for partial oxidation or oxidative dehydrogenation reactions, where undesirable consecutive gas phase reactions leading to total oxidation occur very often. By separating the process stream and the oxidant, the extent of these side-reactions can be drastically diminished. The oxidative coupling of methane is also an example of a reaction where consecutive gas phase total oxidations reduce the selectivity (Roos et al. 1989), and one may take advantage of membrane reactor configuration as has been shown by Omata et al. (1989).

The contact between two reactants can be mediated and controlled by means of a semipermeable membrane. It is well known that for many oxidation reactions subsurface adsorbed species strongly attached to the catalyst surface are responsible for selective products (Kokes and Rennard 1966, Bhattakarya, Nag and Ganguly 1971, Robb and Harriott 1974, Lefferts, van Ommen and Ross 1986). The surface of a dense silver membrane foil through which oxygen diffuses is expected to exhibit higher concentrations of strongly bound oxygen than if the oxygen were adsorbed from the gas phase. A partial oxidation reaction which requires this form of oxygen will proceed more selectively by means of a membrane. The oxidation of ethylene to acetaldehyde is such an example (Gryaznov, Vedernikov and Gul'yanova 1986, Kolosov et al. 1988). The same principle can also be applied to selective hydrogenation reactions, thus incorporating hydrogen having diffused through palladium-based dense membranes (Gryaznov et al. 1976, 1981b, Gryaznov and Smirnov 1977, Gryaznov and Karavanov 1979, Roshan et al. 1983, Sokol'skii, Nogerbekov and Godeleva 1986), or even for other reactions types, such as the water-gas reaction, by means of a palladium membrane through which carbon has been diffused (Ziemecki 1988).

The stoichiometry of the reaction can be regulated. Consider the case where the two reactants of a fast reaction are fed to a porous, inherently catalytic active membrane from opposite sides of the membrane. After having diffused through the membrane they will meet and react, forming in this way a reaction interface. For a fast reaction, penetration of one of the reactants to the opposite side of the membrane can be avoided. Moreover, when a fluctuation of the feed fluxes occurs disturbing the stoichiometry at the reaction interface, the reaction interface will move within the membrane to another position where the stoichiometry is satisfied (self-adjustment; Sloot 1991). This operation mode is of importance for reactions with high conversions where stoichiometry is required whereas the concentrations of both reactants at the exit streams have to be low. Typical examples of such a processes are the desulfurization of gases by the Claus reaction (Sloot 1991) and the reduction of emission gases (de-NO_x reaction).

A membrane reactor offers the possibility of combining two individual processes in the same unit operation. (1) Selective permeation (thus separation) can be coupled directly with the reaction by means of either a catalytically active membrane or of a passive membrane placed next to the reaction zone and forming the wall of the reactor. Equilibrium limited reactions, particularly those liberating hydrogen which is relatively easy to separate, are typical applications of this mode (Ilias and Govind 1989). Since some porous ceramic membranes (e.g. γ-Al$_2$O$_3$) show significant surface diffusion of water, dehydration reactions might also be considered as a serious potential possibility, provided that the stability of the membrane in the presence of water is good. (2) Two reaction processes can be combined and carried out simultaneously in a single-membrane reactor. Consider the reaction chamber to be divided by a membrane selectively permeable to hydrogen. The membrane acts as a catalyst such that hydrogen is split-off from a component supplied from one side (dehydrogenation side), and diffuses through the membrane to be taken up by a component supplied from the other side (hydrogenation side). The removal of hydrogen from the dehydrogenation zone shifts the equilibrium to give an increased yield of the end product, and the diffusion of hydrogen through the membrane gives highly active hydrogen on its surface, so that the hydrogen uptake can be carried out more advantageously. Possible reaction combinations are, e.g. dehydrogenation of alkylaromatic hydrocarbons or olefins and hydrogenation of aromatic hydrocarbons using hydrogen-permeable palladium membranes (Gryaznov 1970, Gryaznov et al. 1970, Itoh and Govind 1989). (3) Multilayer organization of the porous membrane is possible. Successive depositions of thin catalytically active membrane films (e.g. γ-Al$_2$O$_3$/TiO$_2$) result in bi-functional systems suitable for the coupling of consecutive catalytic reactions (Furneaux, Davidson and Ball 1987, Catalytica 1988).

It is advantageous to combine membrane structural and transport properties with the reaction. (1) As a physical substrate, a porous catalytically active membrane causes the catalytic species to be present at a high packing density, dispersion and surface area as compared to the pure phase. (2) Thin catalytic membrane layers with uniform pore size distribution give rise to small, uniform and well controllable residence times. Unstable intermediates can be swept away immediately from the reaction zone. (3) The Knudsen diffusion mechanism which dominates the gas transport through porous catalytically active membranes, where molecule–molecule collisions are not favored, provides conditions that either inhibit gas phase reactions or allow the simultaneous passage of two independent reactant fluxes through the system without mutual interaction. This property might be of importance for an "in-situ" catalyst regeneration, by burning independently the deposited carbon during the course of a dehydrogenation reaction (Zaspalis et al. 1990). Another application is the incorporation of an oxidant in the process for catalyst reactivation, by creating certain adsorbed species on the catalyst surface without causing nonselective reactions (Zaspalis et al. 1990).

7.3. FUNDAMENTAL ASPECTS OF MEMBRANE REACTORS

7.3.1. Separative Membranes

An overview of the decomposition and dehydrogenation reactions that have been investigated using semipermeable membranes for selective permeation of one of the reaction products is given in Tables 7.1 and 7.3, respectively. An overview of the most interesting studies is given in Tables 7.2 and 7.4.

It is evident, that in any membrane reactor operation mode there are important parameters which determine the performance of the process (Shah, Remmen and Chiang 1970). These are: (1) the total and partial pressures on both sides of the membrane, (2) the total and partial pressure differences across the membrane, (3) the diffusion mechanism through the support and the membrane layer (membrane structure), (4) the thickness of the membrane, (5) the reactant configuration (i.e. whether the reactants are supplied from the same or from opposite sides of the membrane, in counter or co-current flow) and (6) the catalyst distribution.

When the membrane performs only a separation function and has no catalytic activity, two membrane properties are of importance, the permeability and the selectivity which is given by the separation factor. In combination with a given reaction, two process parameters are of importance, the ratio of the permeation rate to the reaction rate for the faster permeating component (e.g. a reaction product such as hydrogen in a dehydrogenation reaction) and the separation factors (permselectivities) of all the other components (in particular those of the reactants) relative to the faster permeating gas. These permselectivities can be expressed as the ratios of the permeation rates of

Table 7.1. Membrane Reactor Studies on Decomposition Reactions

Reaction System/Membrane Material	References
Decomposition of Hydrogen Sulfide	
Porous γ-Al_2O_3/MoS_2 membranes	Abe (1987)
Porous γ-Al_2O_3 membranes	Kameyama et al. (1981)
Porous Vycor glass membranes	Kameyama et al. (1979, 1981a, b)
Decomposition of Hydrogen Iodide	
Porous Vycor glass membranes	Itoh et al. (1984)
Nonporous Pd/Ag membranes	Yeheskel, Léger and Courvoisier (1979)
Decomposition of RuO_4 to RuO_2 and O_2	
Porous Fe_2O_3/molecular sieve membranes	Peng, Wang and Zhou (1983)
Decomposition of Water	
Nonporous Pd membranes	Compagnie des Metaux Precieux (1976)

Table 7.2. Summarized Results on Inorganic Membrane Reactors Used for Decomposition Reactions

Membrane Reactor System	Reaction/Catalyst	Results	Reference/Remarks
Porous Vycor glass tube (in double pipe configuration), wall thickness 3 mm, length 600 mm, outer diameter* 15 mm, mean pore diameter 45 Å. Feed enters the reactor at shell side*, permeate at tube side*.	Decomposition of hydrogen sulfide. MoS_2 catalytic pellets packed on the shell side of the reactor.	$T \sim 800°C$, $P_{feed} \sim 4$ bar, $P_{perm} \sim 1$ bar. Conversion of H_2S twice as much as the equilibrium conversion at this temperature.	Kameyama et al. 1981(a, b). The presence of nonseparative viscous flow might have been the reason for low separation factors. Authors state a significant shrinkage of Vycor glass at $T < 800°C$
Porous alumina tube (in double pipe configuration), wall thickness 1 mm, length 200 mm, outer diameter, 15 mm, pore diameter 1020 Å. Feed enters the reactor at tube side, permeate at shell side.	Separation experiments on the system H_2–H_2S. Reaction experiments on the decomposition of hydrogen sulfide over MoS_2 catalysts packed at tube side.	$T \sim 800°C$, $P_{feed} \sim 4$ bar, $P_{perm} \sim 1$ bar. Conversion of H_2S twice as much as the equilibrium conversion at the same temperature.	Kameyama et al. (1981b). Structural stability of Al_2O_3 membrane until 1000°C. Permeability of Al_2O_3 membrane 100 times higher than that of Vycor glass. Authors state that thinner membranes and smaller pores would give better results.
Nonporous palladium/silver tube (23 wt.% silver, in double pipe configuration), wall thickness 0.2 mm, outer diameter 10.8 mm, length 24.3 cm. Feed enters the reactor at tube side, permeate at shell side.	Decomposition of hydrogen iodide. No catalyst present (gas phase decomposition)	$T \sim 500°C$, $P_{feed} \sim 8$ bar, $P_{perm} \sim 1$ bar. The conversion of HI was 4%. In the absence of the membrane the conversion was 0.2%.	Yeheskel, Léger and Courvoisier (1979). In the observed enhanced conversions there might be also a contribution from a catalytic effect of the membrane.

Table 7.2. (Continued)

Membrane Reactor System	Reaction/Catalyst	Results	Reference/Remarks
Three-layer membrane tube (in module configuration, 10 tubes per module, heat exchanger type). *Inner layer*: MoS$_2$ film, thickness 100 μm, mean pore diameter 0.5 μm. *Middle layer*: Cordierite, thickness 1 mm, mean pore diameter 5 μm. *Outer layer*: γ-Al$_2$O$_3$, thickness 50 μm, mean pore diameter 40 Å Feed enters the reactor at tube side, permeate at shell side.	Decomposition of hydrogen sulfide. MoS$_2$ catalyst either in layer form (see inner layer in previous column) or in pellet form (see alternative systems 1 and 2, below)	$T = 800°$C, $P_{feed} \sim 3.8$ bar, $P_{perm} \sim 0.5$ bar. The H$_2$S conversion with MoS$_2$ catalyst in layer form was twice the conversion achieved using alternative system 1 and four times the conversion achieved using alternative system 2.	Abe 1987. Mentioned results were the best among several different configurations which were experimentally tested.
Alternative system 1: Same as above but with MoS$_2$ pellets on tube side instead of MoS$_2$ inner layer	*Alternative system 2*: Tubes consist only of the middle layer (cordierite) with MoS$_2$ pellets packed on tube side		

* The term "tube side" in this and the following tables refers to the inner side of the inner tube of the double pipe configuration shown in Figure 7.1b. Similarly the term "shell side" refers to the outer side of the inner tube (thus the inter-tubular area) of the double pipe configuration shown in Figure 7.1b. The terms "outer" and "inner" diameters refer to the diameters of the membrane tube

Table 7.3. Membrane Reactor Studies on Dehydrogenation Reactions

Reaction System/Membrane Material	References
Dehydrogenation of Cyclohexane to Benzene	
Porous Al_2O_3 membranes	Fleming (1987)
Porous Vycor glass membranes	Shinji et al. (1982), Itoh (1987, Itoh et al. 1988) Sun and Khang (1988)
Nonporous Pd/Ag membranes	Wood (1968), Itoh (1987), Gryaznov (1970)
Dehydrogenation of ethylbenzene to styrene	
Porous Al_2O_3 membranes	Liu et al. (1989), Bitter (1986), Wu et al. (1990)
Dehydrogenation of Ethane to Ethylene	
Porous Al_2O_3 membranes	Furneaux, Davidson and Ball (1987)
Nonporous Pd/Ag membranes	Pfefferie (1966)
Dehydrogenation of Propane to Propene	
Porous Al_2O_3 membranes	Bitter (1986)
Dehydrogenation of 2- or 3-Methyl-1-butene to Isoprene	
Porous Al_2O_3 membranes	Bitter (1986)
Water Gas Shift Reaction	
Nonporous Pd membranes	Kikuchi et al. (1989)
Dehydrogenation of Isopropyl Alcohol to Acetone	
Nonporous Pd membranes	Mikhalenko, Khrapova and Gryaznov (1986)
Dehydrogenation of n-heptane to Toluene and Benzene	
Nonporous Pd/Rh membranes	Gryaznov (1986)
Dehydrogenation of Butene to Butadiene	
Nonporous Pd membranes	Gryaznov et al. (1970)
Dehydrogenation of 1,2-cyclohexanediol	
Nonporous Pd/Cu membranes	Mishchenko et al. (1977)
Steam Reforming of Methane	
Nonporous Pd-alloy membranes	Nazarkina and Kirichenko (1979)

each of the reaction components to the permeation rate of the faster permeating component which is taken as reference.

In the case of dense membranes where only one component (usually a reaction product) can permeate through the membrane, the permselectivities for all the other components are zero, and the extent of the equilibrium shift is determined only by the ratio of the permeation rate to reaction rate for the permeating component. For values of this ratio approximately equal to unity (e.g. for a dehydrogenation reaction this means that the production rate of

Table 7.4. Summarized Results on Inorganic Membrane Reactors Used for Dehydrogenation Reactions

Membrane Reactor System	Reaction/Catalyst	Results	Reference/Remarks
Porous Vycor glass tube (in double pipe configuration), wall thickness 2.4 mm, inner diameter 4.4 mm, length 95 mm, mean pore diameter 40 Å. Feed enters the reactor at tube side, permeate at shell side.	Dehydrogenation of cyclohexane (1) Pt catalyst deposited within the pores of the membrane (Pt 34 wt.%) (2) Pt/Vycor glass pellets packed on tube side (Pt 34 wt.%).	$T \sim 270°C$, $P_{feed} \sim 1.9$ bar, $P_{perm} \sim 0.9$ bar. (1) Conversion of C_6H_{12} 56%, (2) Conversion of C_6H_{12} 40%. The equilibrium conversion of C_6H_{12} at 270°C is 26%.	Sun and Khang (1988). At high space times t (t = catalyst volume/feed flux) the performance is superior than (2) at low space times ($t < 10$ s). Both modes resulted in conversions close to the equilibrium conversion.
Porous Vycor glass tube (in double pipe configuration), wall thickness 1.4 mm, inner diameter 15.9 mm, length 140 mm, mean pore diameter 40 Å. Feed enters the reactor at tube side, permeate at shell side.	Dehydrogenation of cyclohexane Pt/Al$_2$O$_3$ catalytic pellets packed on tube side.	$T \sim 215°C$, $P_{feed} \sim 1$ bar, $P_{perm} \sim 1$ bar. Conversion of C_6H_{12} 80%. The equilibrium conversion of C_6H_{12} at 215°C is 35%.	Shinji, Remmen and Chiang (1970). Authors state that further improvements can be achieved by selecting materials with higher permeability and selectivity.
Porous glass tube (in double pipe configuration), wall thickness 2.7 mm, inner diameter 14.4 mm, length 391 mm, mean pore diameter 40 Å. Feed enters the reactor at tube side, permeate at shell side.	Dehydrogenation of cyclohexane Pt/Al$_2$O$_3$ catalytic pellets (0.5 wt.% Pt) packed on tube side.	$T \sim 180°C$, $P_{feed} \sim 1$ bar, $P_{perm} \sim 1$ bar. Conversion of C_6H_{12} 24%. The equilibrium conversion of C_6H_{12} at 180°C is 10%.	Itoh et al. (1988) Experiments at higher temperatures (until 220°C) also led to similar results.
Porous supported alumina membrane tube (in double pipe configuration), mean pore diameter 40 Å. Feed enters the reactor at tube side, permeate at shell side.	Dehydrogenation of cyclohexane. Pd catalyst deposited within the pores of the membrane.	$T \sim 200-350°C$, $P_{feed} \sim 1$ bar, $P_{perm} \sim 1$ bar. Conversion of C_6H_{12} 80–100%. The equilibrium conversion of C_6H_{12} at 200°C is $<20\%$.	Fleming (1987) Only a few experimental data are provided.

Membrane	Reaction	Conditions and results	Reference
Porous supported multilayer Al_2O_3 membrane tube (in double pipe configuration), inner diameter 0.635 mm, length 250 mm. 1st layer (inner side): Thickness 5 μm, mean pore diameter 40 Å, porosity 50%. 2nd layer: Thickness 30 μm, mean pore diameter 0.2 μm, porosity 35%. 3rd layer: Thickness 50 μm, mean pore diameter 0.8 μm, porosity 40%. Support (outer side): Thickness 1.5–2 mm, mean pore diameter 10–15 μm, porosity 40–45%. Feed enters the reactor at tube side, permeate at shell side.	Dehydrogenation of ethylbenzene to styrene. Fe_2O_3/Al_2O_3 catalyst pellets (occasionally potasium doped) packed on tube side of the reactor.	$T \sim 600\text{–}660°C$, $P_{feed} \sim 1.1\text{–}1.2$ bar, $P_{perm} \sim 1\text{–}1.1$ bar. Conversions of ethylbenzene in the presence of the membrane were 15% higher than those in the absence of the membrane (e.g. 60 and 45%, respectively). The selectivity to styrene was also enhanced in the membrane reactor by $\sim 3\text{–}4\%$ compared to the selectivity in the normal reactor.	Wu et al. (1990). The structural stability of the membrane layer (inner layer) is influenced during the reaction. The mean pore diameter was increased to 60–80 Å, the permeability was increased by $\sim 42\%$ and the separation factor decreased.
Porous supported γ-Al_2O_3 tube (in module configuration), membrane thickness 4–10 μm, mean pore diameter 0.01 μm. Feed enters the reactor at shell side, permeate at tube side.	Dehydrogenation of ethylbenzene to styrene. $Li_{0.5}Fe_{2.4}Cr_{0.1}O_4$ with 12 wt.% K_2O and 3 wt.% V_2O_5 catalytic pellets packed on reactor shell side.	$T \sim 625°C$, $P_{feed} \sim 4$ bar, $P_{perm} \sim 1$ bar. Conversion of ethylbenzene 65.2%, selectivity to styrene 94%. In the absence of membrane the conversion was 50.7%.	Bitter (1986)
Same as above.	Dehydrogenation of propane to propene Cr_2O_3 (20 wt.%)/Al_2O_3 catalytic pellets packed on the shell side of the reactor.	$T \sim 575°C$, $P_{feed} \sim 1$ bar, $P_{perm} \sim 0.25$ bar. Conversion of propane 58.7%, selectivity to propene 90%. In the absence of membrane the conversion was 40.1%.	Same as above

Table 7.4. (Continued)

Membrane Reactor System	Reaction/Catalyst	Results	Reference/Remarks
Same as above.	Dehydrogenation of methyl-1-butene to isoprene. Pt and Sn promoted $ZnAl_2O_4$ catalytic pellets packed on the shell side of the reactor.	$T \sim 550°C$, $P_{feed} \sim 4$ bar, $P_{perm} \sim 1$ bar. Conversion of C_5H_{10} 37.3%, selectivity to C_5H_8 85%. In the absence of membrane the conversion was 27.3%.	Same as above
Dense palladium tube (in double pipe configuration), wall thickness 200 μm, outer diameter 17 mm, length 140 mm. Feed enters the reactor at tube side.	Dehydrogenation of cyclohexane. Pt/Al_2O_3 catalytic pellets (0.5 wt. % Pt) packed at the tube side of the reactor.	$T \sim 200°C$, $P_{feed} \sim 1$ bar, $P_{perm} \sim 1$ bar. Conversion of C_6H_{12} 99.7%. The equilibrium conversion at 200°C was 18.7%.	Itoh (1987) These results are achieved at high feed/purging flow rate ratios.
Microporous glass tube (in double pipe configuration), wall thickness 0.8 mm, outer diameter 10 mm, mean pore diameter 300 nm. Feed enters the reactor at tube side, permeate at shell side.	Water gas shift reaction. Thin dense film (thickness of 20 μm) of Pd catalyst coating the external glass tube surface.	$T \sim 400°C$, $P_{feed} \sim 5$ bar, $P_{perm} \sim 1$ bar. Conversion of CO 100%. The equilibrium conversion was ~ 60%.	Kikuchi et al. (1989) Under the best operating conditions, the rate of hydrogen permeation through the Pd film was evaluated to be of the same order as the rate of production of hydrogen due to the reaction.

hydrogen equals the permeation rate of hydrogen through the membrane), reactions can be driven to completion and 100% conversion can be achieved. This has been theoretically and experimentally demonstrated for the cyclohexane dehydrogenation (Itoh 1987) and for the water-gas shift reaction (Kikuchi et al. 1989), with the use of dense palladium membranes permeable only to hydrogen. Complete conversion has been achieved under conditions where the permeation rate of hydrogen through the membrane was of the same order as the production rate of hydrogen from the reaction, while other reactants or products could not permeate through the membrane. For dense membranes with low permeabilities the attainment of permeation rates which are equal to reaction rates very often implies either very long reactors (large membrane surface to obtain higher permeation fluxes) or high sweep gas fluxes at the permeation* (or separation) side (in order to prevent back-diffusion and to increase the permeation rate) or low reactant fluxes at the reaction side (in order to decrease the production rate) provided that pressure differences across the membrane remain at low levels (0.5–1.5 bar), Kikuchi et al. (1989) reported a hydrogen permeation flux of 0.26 mol-m^{-2}-s^{-1} at 400°C with a pressure difference across the membrane of 0.5 bar and with a membrane thickness of 20 μm. High reaction to separation side flow rate ratios result in low concentrations of the permeating component (e.g. hydrogen) at the separation side, making thus economically infeasible any further recovery of the separated component, in case that this is desirable.

The situation is somewhat different with porous membranes, where the permselectivities for all components do not equal zero but exhibit certain values determined in most cases by the Knudsen law of molecular masses. In general, when porous membranes are used as separators in a membrane reactor next to the catalyst or the reaction zone (Figure 7.2a), it has been shown experimentally (Yamada et al. 1988) and theoretically (Mohan and Govind 1986, 1988a, b, Itoh et al. 1984, 1985) that there is a *maximum equilibrium shift* that can be achieved. On the basis of simple mass balances one can calculate that this maximum depends on, besides the reaction mechanism, the membrane permselectivities (the difference in molecular weights of the components to be separated) and it corresponds to an optimum permeation to reaction-rate ratio for the faster permeating component (which is a reaction product).

Below the optimum value of this permeation to reaction-rate ratio the effect of separation of this product is stronger than the permeation of

---

* The term "separation side" or "permeation side" of the membrane refers to the side of the membrane opposite to the side from which the feed is supplied. For example, for the supported membrane systems shown in Figure 7.2 with the feed entering from the side of the top layer, the separation or permeation side is the side of the support.

reactants and thus a reaction equilibrium shift is obtained. By increasing the permeation to reaction-rate ratio for the faster permeating component up to the optimum value, the conversion is also increased. At values of the permeation to reaction-rate ratio higher than the optimum value, significant permeation of reactants also occurs simultaneously with the products, thus shifting the equilibrium backwards and diminishing the total conversion. Experiments for some common reactions such as the dehydrogenation of cyclohexane or the decomposition of hydrogen sulfide by means of Knudsen diffusion porous membranes placed next to the catalyst, confirm the model predictions (Kameyama et al. 1981a, b, Itoh et al. 1988).

Equilibrium conversions could be increased by a factor ranging from 1.3 to 2.3 depending on the reaction conditions. Theoretical and experimental studies have shown that with the creation of special conditions such as recycling of unconverted reactant or backpermeation of lost reactant, it is possible to achieve conversions even higher than the calculated maximum limits (Mohan and Govind 1988a, Shinji, Misono and Yoneda 1982). The maximum value of the conversion that can be achieved under optimum conditions (thus at the optimum value of the permeation to reaction-rate ratio for the faster permeating product) depends also on the permselectivities of the membrane for the other components. For a given reaction the lower the permselectivities of the reactants (thus the more selective the membrane is for the faster permeating product) the higher the equilibrium shift that can be achieved. If the membrane is permeable only to one component (and the permselectivities to all other components are equal to zero, as is the case for dense membranes) complete conversions can be achieved.

It is obvious from the above discussion that porous and dense membranes form two different cases, each with its own advantages and disadvantages. Dense membranes, (permeable only to one component) operating at optimum conditions, can be used to obtain complete conversions. However, because the permeation rate is low, the reaction rate has also to be kept low. Porous membranes (permeable to all components but at different permselectivities) are limited under optimum conditions to a maximum conversion (which is not 100%) due to the permeation of all the components. The permeation rates through porous membranes are, however, much higher than those through dense membranes and consequently higher reaction rates or smaller reactor volumes are possible.

Microporous membranes (pore radius less than 10 Å) are ideal materials to be used as separators in membrane reactor processes. Microporous membranes also combine the high selectivities to certain components with high permeation rates. The high selectivities mean that maximum conversions (and thus equilibria shifts) higher than those achieved by porous membranes can be attained, while the high permeation rates allow for high reaction rates at the optimum conditions. The high selectivities of the dense membranes and the high permeation rates of the porous membranes can thus be combined perfectly in the structure of the microporous membranes.

In the separation mode, the effect of the membrane thickness on the process is related directly with the above mentioned aspects. The diffusive flow through a dense or a porous membrane can be described by the general formula

$$F \sim D\Delta P^n/L \qquad (7.1)$$

where F is the permeate flux (mol-s^{-1}), D is a constant incorporating the diffusion coefficient (mol-m-bar^{-n}-s^{-1}), ΔP^n is the partial pressure difference function across the membrane ($n = 1$ for porous membranes while $n = 0.5$ for dense palladium membranes), and L is the membrane thickness. From Equation (7.1) it is seen that by decreasing the membrane thickness an increase in permeate flux can be obtained. Thus, thin dense membranes will probably be better than thick ones while for porous membranes there exist an optimum thickness above or below of which no further increase in conversion can be obtained (Yamada et al. 1988; Itoh et al. 1985). This optimum thickness is the one that gives the optimum permeation to reaction-rate ratio discussed above. The effect of the membrane pore size (with pore sizes in the Knudsen permselective regime, i.e. < 100 Å at low pressures), porosity and tortuosity are similar to that of thickness since they are related to the permeation rate via the effective diffusivity. Microporous membranes with very low (but nonzero) permselectivities to the reactants are expected to operate best at optimum thicknesses which are much lower than those of the mesoporous membranes.

Data on palladium dense membranes with some commonly studied reactions such as the dehydrogenation of cyclohexane over Pt/Al$_2$O$_3$ catalysts show values of permeation to reaction-rate ratios much lower than 1, which is the ideal value in the case of dense membranes. From a design point of view and for achieving high conversions, this means either low reactant fluxes or long tubular reactors so that long residence times can be achieved (high Damköhler numbers). A technical challenge is the design of reactors able to provide large surface areas per unit reactor volume. It is obvious that for separative membranes not only the selectivity is of importance but the permeability as well. Gryaznov et al. (1981a), could increase the productivity of a reactor used for dehydrogenation of organic compounds by more than 100 times by replacing the dense palladium foil with a mesoporous glass membrane impregnated with palladium. However, no information concerning the selectivities is provided.

In the last several years research on separation processes with porous membranes have shown that significant separation factors can be achieved via surface diffusion or capillary condensation mechanisms (Asaeda and Du 1986). Gavalas, Megiris and Nam (1989), reported hydrogen separation factors of 300 using silicon dioxide "dense" films chemically deposited in the pores of an alumina microporous support. Uemiya et al. (1988), using an electroless-plating technique, deposited palladium on a porous glass which appeared to be permeable only to hydrogen. Gavalas, Megiris and Nam (1989), as well as Uemiya et al. (1988), report hydrogen permeabilities 10 times higher than the permeabilities achieved with normal dense foils due to the very small thickness of the deposited films. Hydrogen permeation is believed to occur via an activated solution/diffusion mechanism. However, the membrane stability under steam atmosphere or high temperatures appeared in both cases to be questionable. Recent experiments on multilayer adsorption and capillary condensation with alumina membranes (pore diameter 4 nm) showed for a mixture of propylene and nitrogen, separation factors of about 80 (while the Knudsen separation factor of these gases is 1.2) with permeabilities of the order of 4×10^{-5} mol-s^{-1}-Pa^{-1}-m^{-2} (Uhlhorn 1990). Applications of membrane reactors using this mechanism remain a technical challenge. High temperatures require very high pressures for this kind of separations (unfavorable for dehydrogenation reactions and undesirable from a mechanical point of view), or the membrane has to be maintained continuously at much lower temperature than the reaction zone.

Porous ceramic membrane layers are formed on top of macroporous supports, for enhanced mechanical resistance. The flow through the support may consist of contributions due to both Knudsen-diffusion and convective nonseparative flow. Supports with large pores are preferred due to their low resistance to the flow. Supports with high resistance to the flow decrease the effective pressure drop over the membrane separation layer, thus diminishing the separation efficiency of the membrane (van Vuren et al. 1987). For this reason in a membrane reactor it is more effective to place the reaction (catalytic) zone at the top layer side of the membrane while purging at the support side of the membrane.

Purging of the support side of a membrane reactor is also an important process parameter. This can be done either with inerts or with a coupled reaction (Gryaznov 1970, Gryaznov et al. 1970, Itoh 1987, Shinji, Misono and Yoneda 1982) at the side of the support. Fast purging has the advantage of lowering the partial pressures of the permeating substances, maintaining high concentration gradients and thus high diffusion driving forces and permeation rates. Backdiffusion and concentration polarization phenomena due to insufficient permeation rates can also be diminished.

Beside the partial pressure differences between the various gas components, the total pressure on both sides of the membrane is also important. Though mean total pressure does not directly affect the permeation rate in the case of a Knudsen diffusion mechanism, it governs the gas flow through the macroporous support. The flow through the support can be directed either towards or away from the membrane top layer by adjusting the total pressure difference on both sides of the reactor. Directing the flow away from the membrane by applying an overpressure on the top layer side of the membrane, will result in higher separation efficiencies. These will be enhanced more if the support side is at sub-atmospheric pressure (Present and de Bethune 1949). On the other hand, directing the bulk flow from the support side towards the membrane by applying an overpressure on the side of the support, will create conditions of backpermeation (Mohan and Govind 1988a).

It has been mentioned earlier that using porous membranes for product separation during the course of an equilibrium reaction, maximum attainable conversions are limited because of reactant permeation. This is the case where the membrane forms the wall of the reactor in which a catalyst is packed. It has also been mentioned that in this mode equilibrium conversions for some slow reactions could be increased by factors ranging between 1.3 and 2.3. Another important operation mode arises when the membrane is inherently catalytic or when the catalytically active species are placed within the membrane pores (catalytically active membrane as shown in Figure 7.2b and 7.2c). In this case, reaction and separation take place simultaneously and are combined in parallel rather than in series as was the case in the previous mode.

Studies in the literature using catalytically active membranes (Fleming 1987, Sun and Khang 1988) indicate that under certain conditions the calculated maximum attainable conversions can be experimentally exceeded. Abe (1987) (see also Table 7.2) reported an increase in the conversion by a factor of 4, for the decomposition of hydrogen sulfide over catalytically active MoS_2 membrane catalysts, a well-known reaction with high endothermicity and unfavourable thermodynamics up to 1500°C. Fleming (1987) (see also Table 7.4) also reported a conversion four times higher than the equilibrium conversion for the dehydrogenation of cyclohexane using Pd/Al_2O_3 catalytically active membrane. Experimental conversions in both studies cited above were higher than the maximum values predicted by Mohan and Govind (1988a), for the operation mode shown in Figure 7.2a. The important parameter is the space time of the reactant stream in the feed chamber of the reactor (Sun and Khang 1988, see also Table 7.4). At low space times both the passive membrane reactor and the catalytically active membrane reactor are not very efficient in equilibrium shifting. This is related to the high membrane resistance resulting in low flow which limits either sufficient permeation of product or sufficient use of the catalyst which is deposited in the membrane (An analogous phenomenon occurs in conventional catalysis where high Thiele numbers result in low effectiveness factors). At high space times (i.e. lower feed fluxes or longer reactors) both types of membrane reactors are able to shift equilibria with the catalytic membrane reactor being superior to the passive membrane reactor due to the effect of simultaneous reaction and separation. For example, for the platinum-catalyzed dehydrogenation of cyclohexane and for space times of 34 s the passive membrane reactor gave a conversion of 60% which is the maximum attainable as calculated by Mohan and Govind (1988a), while the catalytic membrane reactor gave a conversion of 70% (under these conditions the equilibrium conversion of cyclohexane is 40%). So, the location or the distribution of the catalyst not only plays a significant role in the process but there are strong indications that under conditions of high residence time, it is better to deposit the catalyst within the membrane pores than forming a packed bed in the reactor, adjacent to the membrane.

7.3.2. Nonseparative Porous Membranes

In a less investigated concept, catalytically active membranes have been used to regulate the contact between two reactants for irreversible reactions as previously discussed. For an inherently catalytic porous membrane, the membrane thickness is not only related to the permeation rate but also to the residence time and the number of active sites. Parameters such as purging rate, partial and total pressure differences influence the process in an analogous way as discussed previously. It may be noted at this point that in porous catalytically active membranes where gas transport occurs via Knudsen diffusion, the only way to vary the residence time is to vary the membrane thickness. Increased partial pressure gradients will cause higher fluxes, however, as long as the flow mechanism remains in the Knudsen regime, the time that each molecule will spend in the catalytic membrane remains the same and is independent of the applied driving force contrary to conventional catalytic beds (recall that the Knudsen diffusion coefficient is also independent of pressure). An overview of studies using porous or dense inorganic membranes to mediate and control the contact between the reactants entering the membrane from opposite membrane sides is given in Tables 7.5 and 7.6.

An important aspect concerning catalytically active membrane reactors, is the distribution of the active phase within the membrane system. Modern modification techniques (van Praag et al. 1989, Lin, de Vries and Burggraaf 1989) allow control over the catalyst distribution and preferential deposition of the active phase at different places in the membrane (top layer/support) system. Studies on conventionally used plate-shaped and cylindrically-shaped catalytically active pellets (Vayenas and Pavlou 1987a, b, Dougherty and Verykios 1987) have shown that nonuniformly activated catalysts (catalysts with nonuniform distribution of active sites according to a certain profile)

Table 7.5. Membrane Reactor Studies on Hydrogenation, Oxidation and Other Reaction Types

Reaction System/Membrane Material	References
Dehydrogenation of Isorpropanol to Propene Porous Al_2O_3 membranes	Furneaux, Davidson and Ball (1987)
Oxidative Dehydrogenation of Methanol Porous Al_2O_3 membranes	Zaspalis et al. (1991)
Oxidation of Carbon Monoxide to Carbon Dioxide Porous Al_2O_3 membranes	Furneaux, Davidson and Ball (1987)
Hydrogenation of Ethylene to Ethane Porous Al_2O_3 membranes	Furneaux, Davidson and Ball (1987)
Nonporous Pd-based membranes	Caga, Winterbottom and Harris (1987)
Hydrogenolysis of Ethane to Methane Porous Al_2O_3 membranes	Furneaux, Davidson and Ball (1987)
Claus Reaction for the Desulfurization of Gases Porous Al_2O_3 membranes	Sloot (1991)
Oxidation of Ammonia to Nitrogen Nonporous Ag membranes	Gryaznov, Vedernikov and Gul'yanova (1986)
Oxidative Dehydrogenation of Ethanol to Acetaldehyde Nonporous Ag membranes	Gryaznov, Vedernikov and Gul'yanova (1986)
Oxidative Dehydrogenation of 1-butene to Butadiene Nonporous Zr–Y–Ti-based membranes	Hazbun (1988)
Oxidation of Ethylene to Ethylene oxide Nonporous Zr–Y–Ti-based membranes	Hazbun (1988)
Oxidation of Propylene to Propylene oxide Nonporous Zr–Y–Ti-based membranes	Hazbun (1988)
Oxidative Methane Coupling Nonporous Zr–Y–Ti-based membranes Nonporous PbO/MgO membranes	Hazbun (1988) Omata et al. (1989)
Conversion of Carbon monoxide with Hydrogen Nonporous Pd-based membranes	Gul'yanova et al. (1988)
Hydrogenation of Butadiene Nonporous Pd membranes	Nagamoto and Inoue (1986)
Hydrogenation of Acetylene Nonporous Pd/Ag membranes	Gul'yanova, Gryaznov and Kanizius (1973)
Hydrogenation of Butenes Nonporous Pd/Sb membranes	Gryaznov et al. (1983)

Table 7.5. (Continued)

Reaction System/Membrane Material	References
Hydrogenation of Pentadiene Nonporous Pd/Ru membranes	Mishchenko and Sarylova (1981)
Hydrogenation of Diene Hydrocarbons Nonporous Pd/Ru membranes	Ermilova et al. (1981)
Hydrodemethylation of Methyl and Dimethylnaphthalenes Nonporous Pd/Mo membranes	Lebedeva (1981)
Hydrodealkylation of Dimethylnaphthalenes Nonporous Pd/Ni membranes	Gryaznov, Smirnov and Slin'ko (1976)
Coupling of Hydrogenation and Dehydrogenation Reactions Nonporous Pd-based membranes	Gryaznov (1970), Gryaznov et al. (1970)

exhibit higher effectiveness factors than catalysts with a uniformly distributed active phase. In particular it can be shown (Vayenas and Pavlou 1987a, b) that from all possible catalyst distributions the effectiveness factor for a given mass of catalyst is maximized if all the catalyst is concentrated at a certain position within the pellet. This is a general conclusion and is independent of the reaction kinetics. The optimum position depends on the reaction mechanism and on the reaction and diffusion rates.

A catalytically active membrane reactor can be viewed as a system with such a distribution of the active phase. An effective use of a membrane reactor is then a question of an appropriate choice of process conditions. For example, under isothermal conditions and in the absence of poisoning it is evident that if the reaction rate increases with reactant concentration, high effectiveness factors can be achieved if the reactants are supplied from the catalytically active membrane side, avoiding in this way the diffusion resistance through the support. However, in cases where the reaction rate decreases with increased reactant concentration (i.e. poisoning or for certain ranges of concentrations for reactions with Langmuir–Hinshelwood kinetics) then it might be advantageous to exploit the diffusion resistances by either supplying the reactants from the support side or by depositing the catalyst at the membrane-support interface.

When the reaction between two reactants supplied from opposite membrane sides is very fast then a reaction interface is formed within the inherently catalytic membrane at a position which satisfies the mass balances. Any change in the reactant fluxes will cause a change in the location of the reaction interface which will be re-established at a new position within the

Membrane Reactor System	Reaction/Catalyst	Results	Reference/Remarks
Porous anodized alumina (plate-shaped membranes), mean pore diameter 80–250 nm. Reactants enter the reactor from opposite membrane sides.	Carbon monoxide oxidation, ethane dehydrogenation, ethane hydrogenolysis, ethene hydrogenation. Pt, Mg, Zn catalysts placed either in the pores of the membrane or at the entrance of the membrane pores.	$T \sim 200°C$. No quantitative data are presented concerning conversions and selectivities.	Furneaux, Davidson and Ball (1987). The location of the catalyst had a remarkable influence on the reaction process. The membranes could operate as bifunctional catalysts.
Porous γ-Al_2O_3 supported membranes (plate-shaped), thickness 5–8 μm, mean pore diameter 4–5 nm. Reactants enter the reactor from opposite membrane sides.	Oxidative dehydrogenation of methanol. Silver catalyst deposited in the pores of the membrane (66 wt. % Ag).	$T \sim 250–400°C$, $P_{feed} \sim 1$ bar, $P_{perm} \sim 1$ bar. At 300°C and at 10% total conversion the selectivity to formaldehyde was 36%.	Zaspalis et al. (1991). (1) Enhanced activity of Ag when finely dispersed in membrane matrix. (2) Coke formation can be prevented on-line by feeding oxygen through the membrane.
Porous Al_2O_3 membrane (plate shaped), thickness 3 mm, mean pore diameter 1 μm. Reactants enter the reactor from opposite membrane sides.	Claus process for the desulfurization of gases. γ-Al_2O_3 catalyst deposited in the pores of the membrane.	$T \sim 250°C$, $P_{feed} \sim 1$ bar, $P_{perm} \sim 1$ bar. 100% conversion within the membrane (fast reaction).	Sloot (1991). Self-control of the feed of reactants (according to stoichiometry) and permeation of one of the reactants to the opposing membrane side is prevented.
Calcia or yttria or magnesia stabilized zirconia tube externally coated with a film Zr:Y:Ti (87:12:1 mol%) using ECVD technique (in double pipe configuration), film thickness 2–10 μm, tube thickness 2 mm, inner diameter 1 cm, length 15 cm.	Ethylene oxidation to ethylene oxide. Silver catalyst placed in the pores of the tube (8 wt. % Ag).	$T \sim 250–400°C$, $P_{feed} \sim 1–2$ bar. Conversion of ethylene 10%. Selectivity to oxide 75%.	Hazbun (1988)
	Propylene oxidation to propylene oxide. Silver catalyst placed in the pores of the tube (8 wt. % Ag).	$T \sim 300–500°C$, $P_{feed} \sim 1–2$ bar. Conversion of propylene 15%. Selectivity to oxide 30%.	

Table 7.6. (Continued)

Membrane Reactor System	Reaction/Catalyst	Results	Reference/Remarks
Feed enters the reactor at tube side, oxygen at shell side.	Oxidative dehydrogenation of 1-butene to butadiene. $W_3Sb_2O_3$ catalyst placed in the pores of the tube.	$T \sim 462°C$. Conversion 30%, Selectivity 92%. $T = 505°C$. Conversion 57%, Selectivity 88%.	
	Oxidative methane coupling. $Mn_8Ca_bZr_cO_x$ catalyst placed in the pores of the tube.	$T \sim 850°C$. Conversion 30%. Selectivity to C_2 products 60%	
	Ethylene and ethane synthesis from methane. Li/MgO (3 wt. % Li) catalyst placed in the pores of the tube.	$T \sim 700–750°C$. Conversion 40%. Selectivity to C_2 products 55%.	
Porous alumina tube externally coated with a MgO/PbO dense film (in double pipe configuration), tube thickness 2.5 mm, outer diameter 4 mm, mean pore diameter 50 nm, active film-coated length 30 mm. Feed enters the reactor at shell side, oxygen at tube side.	Oxidative methane coupling, PbO/MgO catalyst in thin film form (see previous column).	$T \sim 750°C$, $P_{feed} \sim 1$ bar. Conversion of methane $< 2\%$. Selectivity to C_2 products $> 97\%$.	Omata et al. (1989). The methane conversion is not given. Reported results are calculated from permeability data.
Nonporous silver membrane tube (99.99 wt. % Ag), (in double pipe configuration), thickness 100 μm. Feed enters the reactor at shell side, oxygen at tube side.	Oxidation of ammonia. Silver catalyst in membrane form (see previous column).	$T \sim 250–380°C$. The yield of nitrogen was 40%, the yield of nitrogen monoxide was 25%.	Gryaznov, Vedernikov and Gul'yanova (1986)
	Oxidation of ethanol to acetaldehyde. Silver catalyst in membrane form (see previous	$T \sim 250–380°C$. The yield of acetaldehyde was 83%. The yield with bulk powdered silver	

membrane. Then, permeation of one of the reactants to the opposite membrane side can be avoided. This operation mode has been experimentally tested for the Claus process where hydrogen sulfide and sulfur dioxide, entering from opposite membrane sides, react rapidly and reversibly to produce sulfur and water (Sloot, 1991). Calculations have shown that inclusion of convective flow and external mass transfer resistance for most cases does not prevent the shifting in reaction zone mechanism (Sloot 1991). This system does not rely on the membrane selectivity and open structure membranes can be used thus allowing high fluxes, provided that the pore size and the pressure difference across the membrane are such to allow the reaction zone to be in the membrane. It has been shown that with a pressure difference across the membrane of 1 bar and a membrane pore diameter of 1 μm, a porosity of about 47% and a thickness of 3 mm, the reaction zone remains in the membrane.

Though some membrane applications can take advantage of larger pores, there are certain nonseparative operation modes that require small pores and the flow through the membrane in the Knudsen regime. This is the case where one of the streams (e.g. oxygen) is used to regenerate the membrane catalyst deactivated by carbon deposition that takes place during the reaction. The Knudsen diffusion mechanism provides conditions of minimal interaction between the oxidant and the process stream in the membrane; the oxidant can thus be effectively used for burning out the deposited carbon without influencing the reaction. Catalyst lifetimes could be significantly extended by employing this principle and using low oxygen partial pressures. (Zaspalis et al. 1990).

7.3.3. Nonseparative Dense Oxide Membranes

Besides metallic membranes through which hydrogen (Gryaznov et al. 1976, 1981a, b, Gryaznov and Smirnov 1977, Gryaznov 1986, Uemiya et al. 1988) or oxygen (Gryaznov 1986, Kolosov et al. 1988, Itoh and Govind 1989) diffuses, dense ceramic materials too, which conduct oxygen or hydrogen ions, can be used as catalytically active membranes. An example is the lanthanum oxide stabilized bismuth oxide in the oxidative dehydromerization of ethylene (di Cosimo, Burrington and Grasselli 1986). The bismuth oxide membrane acts both as a catalyst and as a separator of the propylene feed from O_2 to prevent over-oxidation to CO_2. Recently, oxidation reactions have been studied by means of dense ceramic films of yttria-stabilized zirconia through which oxygen was diffused from one side, to the catalyst on the other side of the membrane and in contact with the process stream (Hazbun 1988; see also Table 7.6). Selective oxidation of ethylene to ethylene oxide with a selectivity of 75% was obtained. Other reactions reported were oxidation of propylene, oxidative dehydrogenations of olefins and oxidative methane coupling. The oxygen species used in these reactions originate from the lattice of the solid material and, as is the case with metallic membranes, have a character different from gaseous oxygen species.

7.4. ASSESSMENT OF COMMERCIAL POSSIBILITIES

Inorganic membrane reactors offer the possibility to combine reaction and separation in a single stage at high temperatures and to overcome the equilibrium limitations experienced in conventional reactor configurations. They also offer the possibility for an accurate and controlled reactant supply to the reaction zone. Such attractive features can be advantageously utilized in a number of potential commercial opportunities. These include dehydrogenations, hydrogenations, oxidative dehydrogenations, oxidations and catalytic decomposition reactions. However, to be cost effective, significant technological advances and improvements will be required to solve several key issues. These are discussed below.

The smallest pore diameter of the present state-of-the-art commercial microporous inorganic membranes is about 3–4 nm (e.g. Membralox® alumina membranes). Although these have a relatively thin ($\sim$ 2–5 μm) crack-free permselective layer with uniform porosity (as high as 50%), the pore diameter is not low enough to obtain a high separation efficiency. This is due to the fact that for such mesoporous membranes the separation of gases occurs through the Knudsen diffusion mechanism with ideal separation factors below 10 for most gaseous mixtures. For instance, for the catalytic dehydrogenation of ethylbenzene to styrene in a membrane reactor, the maximum separation factor for the separation of H_2 from ethylbenzene or styrene is only about 7 (Liu et al. 1989, Wu et al. 1990). This suggests that in order to achieve efficient gas or vapor separations, other separation mechanisms such as activated microporous diffusion, molecular sieving or surface diffusion, multilayer adsorption and capillary condensation (Uhlhorn et al. 1989b, Gavalas, Megiris and Nam 1989) will be necessary. Ion-conductive thin oxide films made by vapor deposition techniques might also be considered as a serious possibility in membrane applications (Lin, de Vries and Burggraaf 1989).

Such membrane structures are under development but it will require considerable effort to overcome formidable technical challenges on the road to commercialization. A large number of commercially significant opportunities involve (1) high operating temperatures (up to 1000°C such as for the catalytic decomposition of H_2S) under corrosive environments (2) presence of large quantities of water vapor which generally has a destabilizing influence on the membrane structure (Nam and Gavalas 1989) or (3) endothermic reactions such as the dehydrogenation of ethylbenzene (Wu et al. 1990) or exothermic reactions such as hydrogenation of 1,3-cyclooctadiene or naphthalene (Ermilova et al. 1981, Ermilova 1981) and cyclopentadiene (Gryaznov and Karavanov 1979). Here, the mass and heat transfer considerations are of crucial importance (Mohan and Govind 1988c) for the reactor design and have to be coupled with further material and system development.

In addition, challenges in the area of reactor engineering and module development are equally formidable. These involve two major issues. The first is related to membrane module and reactor design in relation to the development of effective seals under high temperature and high pressure (in some applications such as coal off-gas clean-up) operating conditions to withstand repeated thermal cycling. In addition to high-temperature seals, material compatibility of housing and peripherals which are also subject to thermal cycling and accompanying thermal expansion and contraction effect has to be dealt with.

The other major issue in reactor design concerns catalyst deactivation and membrane fouling. Both contribute to loss of reactor productivity. Development of commercially viable processes using inorganic membrane reactors will only be possible if such barriers are overcome. These subjects will receive greater attention as current R&D efforts expand beyond laboratory scale evaluations into field demonstrations.

REFERENCES

Abe, F. 1987. Porous membranes for use in reaction processes. European Patent. Appl. 0,228,885A2.

Armor, J. N. 1989. Catalysis with permselective inorganic membranes. *Appl. Catal.* 49: 1–25.

Anderson, M. A., M. J. Gieselmann and Q. Xu. 1988. Titania and alumina ceramic membranes. *J. Membrane Science* 39: 243–258.

Asaeda, M. and L. D. Du. 1986. Separation of alcohol/water gaseous mixtures by thin ceramic membrane. *J. Chem. Eng. Japan* 19(1): 72–77.

Bitter, J. G. A. 1986. Dehydrogenation using porous inorganic membranes. British Patent Appl. 8,629,135.

Bhattakarya, S. K., N. K. Nag and N. D. Ganguly. 1971. Kinetics of vapor phase oxidation of methanol on reduced silver catalyst. *J. Catal.* 23: 158–167

Burggraaf, A. J. and K. Keizer. 1991. Synthesis of Inorganic Membranes. In *Inorganic Membranes; Synthesis Characteristics and Applications*, Eds R. Bhave, van Nostrand Reinhold, New York (Chapter 2).

Caga, I. T., J. M. Winterbottom and I. R. Harris. 1987. Pd-based diffusion membranes as ethylene hydrogenation catalysts. *Inorg. Chim. Acta* 140: 53–55

Catalytica study. 1988. Catalytic membrane reactors: concepts and applications. 4187MR.

Compagnie des Metaux Precieux. 1976. Hydrogen from water. French Patent Appl. 2,302,273A1.

Cussler, E. D. 1988. Microporous membrane trickle bed reactor. European Patent Appl. 0,293,186A2.

di Cosimo, R., J. D. Burrington and R. K. Grasselli. 1986. Oxidative dehydrogenation of propylene over Bi_2O_3-La_2O_3 oxide ion conductive catalysts. *J. Catal.* 102: 234–239.

Dellefield, R. J. 1988. High-temperature applications of inorganic membranes. Presented at the AIChE National Meeting, session 2f: separations by inorganic membranes 21–24 August, Denver, Colorado.

Dougherty, R. C. and X. E. Verykios. 1987. Nonuniformly activated catalysts. *Catal. Rev. Sci. Eng.* 29(1): 101–150.

Ermilova, M. M. 1981. Selective hydrogenation of nahpthalene on membrane catalysts. *Met. i Kak Membran. Katal. M.* 101–111.

Ermilova, M. M., N. V. Orekhova, L. D. Gogua and L. S. Morosova. 1981. Selective hydrogenation of diene hydrocarbons on a palladium-ruthenium membrane catalyst. *Met. i Kak Membran. Katal. M.* 82–100.

Fleming, H. L. 1987. Latest developments in inorganic membranes. Presented at BBC membrane planning conf. 20–22 October, Cambridge, MA.

Furneaux, R. C., A. P. Davidson and M. D. Ball. 1987. Porous anodic aluminum oxide membrane catalyst support. European Patent Appl. 0,244,970A1.

Furneaux, R. C. and M. C. Thornton. 1988. Porous "ceramic" membranes produced from anodizing aluminium. Brit. Cer. Proc., *Advanced Ceramics in Chemical Process Engineering*, Eds. B. C. H. Steele and D. P. Thompson, vol. 43, pp. 93–101.

Gavalas, G. R., C. E. Megiris and S. W. Nam. 1989. Deposition of H_2-permselective SiO_2 films. *Chem. Eng. Sci.* 44(9): 1829–1835.

Gryaznov, V. M. 1970. Simultaneous contacting of catalytic processes which are associated with the generation and adsorption of hydrogen. German Patent Appl. 1,925,439.

Gryaznov, V. M., V. S. Smirnov, L. K. Ivanova and A. P. Mishchenko. 1970. Coupling of reactions resulting from hydrogen transfer through the catalyst. *Dokl. Akad. Nauk SSSR.* 190(1): 144–147.

Gryaznov, V. M., V. S. Smirnov and M. G. Slin'ko. 1976. Binary palladium alloys as selective membrane catalysts. *Proc 6th Intl. Cong. Catal.*, Eds: G. C. Bond, P. B. Wells and F. C. Tompkins, vol. 2, pp. 894–902.

Gryaznov, V. M. and V. S. Smirnov. 1977. Selective hydrogenation on membrane catalysts. *Kinet. and Catal.* 18(3): 485–486.

Gryaznov, V. M. and A. N. Karavanov. 1979. Hydrogenation and dehydrogenation of organic compounds on membrane catalysts (review). *Khim.-Farm. Zh.* 13(7): 74–78.

Gryaznov, V. M., A. N. Karavanov, O. K. Krusil'nikova, G. L. Chernysova and A. V. Patrikeev. 1981a. Palladium on mesoporous glass as a catalyst for the hydrogenation of unsaturated compounds. *Izv. Akad. Nauk. SSSR, Ser. Khim.* 7: 1663–1666.

Gryaznov, V. M., V. S. Smirnov and M. G. Slin'ko. 1981b. The development of catalysis by hydrogen porous membranes. *Stud. Surf. Sci. Catal.* 7: 224–234.

Gryaznov, V. M., N. M. Ermilova, N. V. Orekhova and N. A. Makhota. 1983. *Heter. Catal.* 5(1): 225.

Gryaznov, V. M. 1986. Surface catalytic properties and hydrogen diffusion in palladium alloy membranes. *Zeits. Für Phys. Chem. Neue Folge* 147: 123–132.

Gryaznov, V. M., V. I. Vedernikov and S. G. Gul'yanova. 1986. Participation of oxygen, having diffused through a silver membrane catalyst, in heterogeneous oxidation processes. *Kinet. and Catal.* 27(1): 129–133.

Guizard C., F. Legault, N. Idrissi, A. Larbot, L. Cot and G. Gavach. 1989. Electronically conductive mineral membranes designed for electro-ultrafiltration. *J. Membrane Science* 41: 127–142.

Gul'yanova, S. G., V. M. Gryaznov and S. Kanizius. 1973. Selective hydrogenation of acetylene on a palladium-silver membrane catalyst. *Analiz Soverm. Zadach u Tech. Naukakh* 172.

Gul'yanova, O. S., Y. M. Serov, S. G. Gul'yanova and V. M. Gryaznov. 1988. Conversion of carbon monoxide on membrane catalysts of palladium alloys; Reaction between CO and H_2 on binary palladium alloys with ruthenium and nickel. *Kinet. and Catal.* 29(4): 728–731.

Haggin, J. 1988. New generation of membranes developed for industrial separations. *Chem. Eng. News* June 6: 7–16.

Hazbun, E. A. 1988. Ceramic membranes for hydrocarbon conversion. U.S. Patent 4,791,079.

Hsieh, P. 1988. Inorganic membranes. *A.I.Ch.E. Symp. Ser.* 84(261): 1–18.

Hsieh, P. 1989. Inorganic membrane reactors; a review. *A.I.Ch.E. Symp. Ser.* 85(268): 53–67.

Hurly, P. 1987. New filters clean up in new markets. *High Technology* 21–24.

Ilias, S. and R. Govind. 1989. Development of high temperature membranes for membrane reactor: an overview. *A.I.Ch.E. Symp. Ser.* 85(268): 18–25.

Itoh, N., Y. Shindo, T. Hakuta and H. Yoshitome. 1984. Enhanced catalytic decomposition of HI by using a microporous membrane. *Int. J. Hydrogen Energy.* 9(10): 835–839.

Itoh, N., Y. Shindo, K. Haraya, K. Obata, T. Hakuta and H. Yoshitome. 1985. Simulation of a reaction accompanied by separation. *Int. Chem. Eng.* 25(1): 138–142.

Itoh, N. 1987. A membrane reactor using palladium. *A.I.Ch.E. J.* 33(9): 1576–1578.

Itoh, N., Y. Shindo, K. Haraya and T. Hakuta. 1988. A membrane reactor using microporous glass for shifting equilibrium of cyclohexane dehydrogenation. *J. Chem. Eng. Japan* 21(4): 399–404.

Itoh, N. and R. Govind. 1989. Development of a novel oxidative palladium membrane reactor. *A.I.Ch.E. Symp. Ser.* 85(268): 10–17.

Kameyama, T., M. Dokiya, K. Fukuda and Y. Kotera. 1979. Differential permeation of hydrogen sulfide through a microporous vycor-type glass membrane in the separation system of hydrogen and hydrogen sulfide. *Separ. Sci. Technol.* 14(10): 953–957.

Kameyama, T., M. Dokiya, M. Fujishige, H. Yokohawa and K. Fukuda. 1981a. Possibility for effective production of hydrogen from hydrogen sulfide by means of a porous vycor glass membrane. *Ind. Eng. Chem. Fundam.* 20(1): 97–99.

Kameyama, T., K. Fukuda, M. Fujishige, H. Yokohawa and M. Dokiya. 1981b. Production of hydrogen from hydrogen sulfide by means of selective diffusion membranes. *Adv. Hydrogen Energy Prog.* 2: 569–579.

Keizer, K. and A. J. Burggraaf. 1988. Porous ceramic materials in membrane applications. *Sci. Cer.* 14: 83–93.

Kikuchi, E., S. Uemiya, N. Sato, H. Inoue, H. Ando and T. Matsuda. 1989. Membrane reactor using microporous glass-supported thin film of palladium; Application to the water gas shift reaction. *Chem. Lett.* 3: 489–492.

Kokes, R. J. and J. R. J. Rennard. 1966. Hydrogenation of ethylene and propylene over palladium hydride. *J. Phys. Chem.* 70, 2543–2547.

Kolosov, E. N., N. I. Starkovskii, S. G. Gul'yanova and V. M. Gryaznov. 1988. Oxygen permeability of thin silver membranes; Effect of the adsorption of benzene on the oxygen transfer process. *Russian J. Phys. Chem.* 62(5): 661–663.

Lakshminarayanaiah, N. 1969. Transport phenomena in membranes. Academic Press. Orlando, FL.

Lebedeva, V. I. 1981. Hydrodemethylation of methyl and dimethylnapthalenes on membrane catalysts from binary palladium alloys. *Met i Splavy Kak Membran. Katalyzatory M.* 112–116.

Lee, E. K. 1987. Membranes: synthesis, applications. *Encycl. of Phys. Sci. Technol.* 8: 21–55.

Leenaars, A. F. M., A. J. Burggraaf and K. Keizer. 1987. Process for the production of crack-free semi-permeable inorganic membranes. U.S. Patent 4,711,719.

Lefferts, L., J. G. van Ommen and J. R. H. Ross. 1986. The oxidative dehydrogenation of methanol to formaldehyde over silver catalysts in relation to the oxygen-silver interaction. *Appl. Catal.* 23: 385–402.

Lin, Y. S., K. J. de Vries and A. J. Burggraaf. 1989. CVD modification of ceramic membranes: simulation and preliminary results. J. de Phys. Coll. de Phys., *Proc. 7th Europ. Conf. on Chemical Vapor Deposition* vol. C5, p. 861.

Liu, Y., A. G. Dixon, Y. H. Ma and W. R. Moser. 1989. Permeation of ethylbenzene and hydrogen through untreated and catalytically-treated alumina membranes. Presented at the 198th ACS National Meeting division of Petrol. Chemical Symposium on New Catalytic Materials, 10–15 September, Miami, FL.

Michaels, A. S. 1968. New separation technique for the CPI. *Chem. Eng. Prog.* 64: 31–43.

Mikhalenko, N. N., E. V. Khrapova and V. M. Gryaznov. 1986. Influence of hydrogen on the dehydrogenation of isopropyl alcohol in the presence of a palladium membrane catalyst. *Kinet. and Catal.* 27(1): 125–128.

Mishchenko, A. P., M. E. Sarylova, V. M. Gryaznov, V. S. Smirnov, N. R. Roshan, V. P. Polyakova and E. M. Savitskii. 1977. Hydrogen permeability and catalytic activity of membranes made of palladium-copper alloys in relation to the dehydrogenation of 1,2-cyclohexanediol. *Izv. Akad. Nauk SSSR, Ser. Khim.* 7: 1620–1622.

Mishchenko, A. P. and M. E. Sarylova. 1981. Hydrogen permeability and catalytic activity of a membrane catalyst from a palladium alloy containing 6% ruthenium in relation to hydrogenation of 1,3-pentadiene. *Met. i Splavy Membrane Kak. Katalyz. M.* 75–81.

Mohan, K. and R. Govind. 1986. Analysis of a cocurrent membrane reactor. *A.I.Ch.E. J.* 32(12): 2083–2086.

Mohan, K. and R. Govind. 1988a. Analysis of equilibrium shift in isothermal reactors with a permselective wall. *A.I.Ch.E. J.* 34(9): 1493–1503.

Mohan, K. and R. Govind. 1988b. Studies on a membrane reactor. *Separ. Sci. Technol.* 23(13): 1715–1733.

Mohan, K. and R. Govind. 1988c. Effect of temperature on equilibrium shift in reactors with a permeselective wall. *Ind. Eng. Chem. Res.* 27(11): 2064–2070.

Nagamoto, H. and H. Inoue. 1985. A reactor with catalytic membrane permeated by hydrogen. *Chem. Eng. Commun.* 34: 315–323.

Nagamoto, H. and H. Inoue. 1986. The hydrogenation of 1,3-butadiene over a palladium membrane. *Bull. Chem. Soc. Japan* 59: 3935–3939.

Nam, S. W. and G. R. Gavalas. 1989. Stability of H_2-permselective SiO_2 films formed by chemical vapor deposition. *A.I.Ch.E. Symp. Ser.* 85(268): 68–74.

Nazarkina, E. B. and N. A. Kirichenko: 1979. Improvement in the steam catalytic conversion of methane by hydrogen liberation via palladium membranes *Khim. Tekhnol. Topl. Masel.* 3: 5–10.

Omata, K., S. Hashimito, H. Tominaga and K. Fujimoto. 1989. Oxidative coupling of methane using a membrane reactor. *Appl. Catal.* 52(L1).

Peng, N., F. Wang and D. Zhou. 1983. Preparation of a filter medium for removing ruthenium and evaluation of its performance. *Fushe Fanghu* (Radiation Protection), 3(2): 154–158.

Pfefferie, W. C. 1966. U.S. Patent Appl. 3,290,406.

van Praag, W. P., V. T. Zaspalis, K. Keizer, J. G. van Ommen, J. R. H. Ross and A. J. Burggraaf. 1989. Preparation, modification and microporous structure of alumina and titania ceramic membrane systems. *Proc. 1st Europ. Ceram. Soc. Conf.* Eds. G. de With, R. A. Terpstra and R. Metselaar, vol. 3 pp. 605–609.

Present, R. D. and A. J. deBethune. 1949. Separation of a gas mixture through a long tube at low pressure *Phys. Rev.* 75(7): 1050–1061.

Raymont, M. E. D. 1975. Make hydrogen from hydrogen sulfide. *Hydroc. Proc.* 54(7): 139–142.

Robb, D. A. and P. Harriott. 1974. The kinetics of methanol oxidation on a supported silver catalyst. *J. Catal.* 35: 176–183.

Roos, J. A., S. J. Korf, J. J. P. Biermann, J. G. van Ommen and J. R. H. Ross. 1989. Oxidative coupling of methane, the effect of gas composition and process conditions. *Proc. 2nd Europ. Workshop Meeting, New Developments in Selective Oxidation.* Rimini, Italy.

Shah, Y. T., T. Remmen and S. H. Chiang. 1970. A note on isothermal permeable wall plug flow reactor. *Chem. Eng. Sci.* 25: 1947–1948.

Shinji, O., M. Misono and Y. Yoneda. 1982. The dehydrogenation of cyclohexane by the use of a porous-glass reactor. *Bull. Chem. Soc. Japan* 55(9): 2760–2764.

Sloot, H. 1991. A non-permselective membrane reactor for catalytic gas phase reactions. Thesis, University of Twente, Enschede.

Sokol'skii, D. V., B. Y. Nogerbekov and N. N. Gudeleva. 1986. Investigation of the activity of a palladium/glass membrane in catalytic hydrogenation reactions. *Sov. Electrochem.* 22(9): 1227–1229.

Sun, Y. M. and Khang, S. J. 1988. Catalytic membrane for simultaneous chemical reaction and separation applied to a dehydrogenation reaction. *Ind. Chem. Eng. Res.* 27(7): 1136–1142.

Suzuki, H. 1987. Composite membrane having a surface layer of an ultrathin film of cage-shaped zeolite and processes for production thereof. U.S. Patent 4,699,892.

Suzuki, F., K. Onozato and Y. Kurokawa. 1987. Gas permeability of a porous alumina membrane prepared by the sol–gel process (aluminium iso-propoxide). *J. Non-Crystl. Solids* 94: 160–162.

Uemiya, S., Y. Kude, K. Sugino, T. Matsuda and E. Kikuchi. 1988. A palladium/porous-glass composite membrane for hydrogen separation. *Chem. Lett.* 10: 1679–1690.

Uhlhorn, R. J. R., M. H. B. J. Huis in't Veld, K. Keizer and A. J. Burggraaf. 1989a. Theory and experiments on transport of condensable gases in microporous ceramic membrane systems. *Proc. 1st Intl. Cong. Inorganic Membrane*, 3–6 July, 323–328, Montpellier.

Uhlhorn, R. J. R., M. H. B. J. Huis in't Veld, K. Keizer and A. J. Burggraaf. 1989b. High permselectivities of microporous silica-modified γ-alumina membranes. *J. Mat. Sci. Lett.* 8: 1135–1138.

Uhlhorn, R. J. R., Ceramic membranes for gas separation; synthesis and transport properties, Thesis, University of Twente, Enschede.

Vayenas, C. G. and S. Pavlou. 1987a. Optimal catalyst distribution and generalized effectiveness factors in pellets: single reactions with arbitrary kinetics. *Chem. Eng. Sci.* 42(11): 2633–2645.

Vayenas, C. G. and S. Pavlou. 1987b. Optimal catalyst distribution for selectivity maximization in pellets: parallel and consecutive reactions. *Chem. Eng. Sci.* 42(7): 1655–1666.

van Vuren, R. J., B. C. Bonekamp, K. Keizer, R. J. R. Uhlhorn, H. J. Veringa and A. J. Burggraaf. 1987. Formation of ceramic alumina membrane for gas separation. *High Tech Ceramics*, ed. P. Vincenzini, p. 2235–2245.

Wood, B. J. 1968. Dehydrogenation of cyclohexane on a hydrogen-porous membrane. *J. Catal.* 11: 30–34.

Wu, J. C. S., T. E. Gerdes, J. L. Pszczolkowski, R. R. Bhave and P. K. T. Liu. 1990. Dehydrogenation of ethylbenzene to styrene using commercial ceramic membranes as reactors. *Separ. Sci. Technol.* 1990. 25(13–15): 1489–1510.

Yamada, M., K. Fugii, H. Haru and K. Itabashi. 1988. Preparation and catalytic properties of special alumina membrane formed by anodic oxidation of aluminum. *Proc. 9th Intl. Cong. Catal.* 1945–1951.

Yeheskel, J., D. Léger and P. Courvoisier. 1979. Thermal decomposition of hydroiodic acid and hydrogen separation. *Adv. Hydrogen Energy.* 2(1): 569–594.

Zaspalis, V. T., W. van Praag, K. Keizer, J. R. H. Ross and A. J. Burggraaf, 1990. Modified alumina membranes as active materials in catalytic processes. *Proc. 1st Intl. Cong. Inorganic Membrane*, 3–6 July, 367–372, Montpellier.

Zaspalis, V. T., K. van Praag, J. G. van Ommen, J. R. H. Ross and A. J. Burggraaf. 1991. The reactions of methanol over alumina catalytically active membranes. Accepted for publication. *Appl. Catalysis.*

8. Inorganic Membranes in Food and Biotechnology Applications

R. R. BHAVE

Alcoa Separations Technology, Inc., Warrendale, PA

J. GUIBAUD

Societe des Ceramiques Techniques, Tarbes

B. TABODO DE LA FUENTE

University of Montpellier, Montpellier

AND

V. VENKATARAMAN*

U.S. Department of Energy, Morgantown

8.1. INTRODUCTION

In the field of food and dairy industry applications, inorganic membranes are now used in a wide range of processes, especially in microfiltration and ultrafiltration. This growth in application is believed to be the result of the availability of high-quality membrane products and due to the constant evolution of membrane technology.

In the beginning (1970s), membrane processes were utilized in the manufacture of cheese and whey protein concentrate. The sphere of application then expanded to milk protein standardization, manufacture of lactic curds, fresh cream cheese by ultrafiltration and the separation of proteins and peptides. Some membrane applications with a significant potential for commercialization include bacterial epuration, manufacture of skimmed milk and pre-concentration of milk at the farm jointly with thermalization.

---

* Formerly with Norton and Millipore

The progress of inorganic membrane technology in food and biotechnology applications occurred in three major areas:

1. Substantial improvements in the separation quality, selectivity and product throughput
2. Availability of materials with high resistance to pH, temperature, solvents and efficient membrane cleaning solutions
3. Advancements in processing equipment

Cross-flow filtration applications can be classified into two principal categories based on the product type, desired separation or retention characteristics: (1) concentration of soluble molecules and suspended solids (particles) and (2) clarification by removing only suspended solids (bacteria, fat or oily particles) to produce clear solutions (milk, beverages, water, metabolites).

The rapid growth in the published literature provided membrane users with new knowledge about the mechanisms at work and descriptions of various problem-solving techniques or approaches. As a result, the membrane market has experienced a phenomenal growth. For the dairy industry alone, the worldwide membrane usage increased from about $300\,m^2$ (of installed surface) in 1971 to approximately $180,000\,m^2$ in 1989. Inorganic membranes enjoy about 10% of market share in the food and dairy industry.

8.2. APPLICATIONS OF INORGANIC MEMBRANES IN THE DAIRY INDUSTRY

8.2.1. Microfiltration of Milk for Bacteria Removal

Although microfiltration has been used in industrial practice for many decades, there are several technical problems and challenges that are yet to be overcome. A detailed understanding of the interaction between the membrane and process feed components is still lacking for many applications in the dairy industry and is further complicated by the significant influence of hydraulic parameters on filtrate flux and fouling. Bacteria removal by microfiltration of milk is used as the case in point.

Microfilters of $0.2\,\mu m$ pore diameter have been successfully utilized to remove bacteria from aqueous solutions such as fermentation media and in sterile filtrations. However, the performance of such filters can be very different in milk filtration due to many complicating factors. One major problem is the size of fat globules (average diameter $1-8\,\mu m$) and protein (casein) micelles ranging from $0.025\,\mu m$ up to as large as $0.3\,\mu m$ (Muir and Banks 1985, Piot et al. 1987). These are comparable to the size of bacteria typically encountered (e.g. $0.2-0.5\,\mu m$ and higher). This situation, as may be visualized, can result in rapid fouling of the membrane. Cross-flow filtration offers the potential to deal with this problem by operating at higher cross-flow velocities to reduce fouling in a feed and bleed operating configuration (Malmberg and Holm 1988).

A typical case using cross-flow microfiltration to produce low-bacteria skimmed milk (also known as the cold pasteurization process) is schematically shown in Figure 8.1. It was shown that by using a microfilter with a pore diameter in the range 1–1.5 μm, more than 99.6% bacteria were retained without significant flux decline (Malmberg and Holm 1988). The performance of the membrane filter and the properties of the product (permeate) are influenced not only by the average values of operational parameters such as transmembrane pressure, ΔP_T, cross-flow velocity, pore diameter, temper-

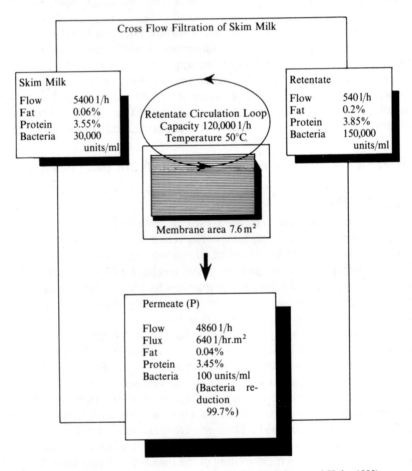

Figure 8.1. Cross-flow filtration of skimmed milk (Malmberg and Holm 1988).

ature, concentration factor, but also by the distribution of these values over the module length and duration of operation.

Two such parameters are the pore distribution of membranes and ΔP_T. Ceramic membranes, due to their narrow pore size distribution, offer an inherent solution to the problem of pore plugging attributable to distribution of pore sizes in the separating barrier. The problem of membrane fouling and consequential decrease in flux as a result of variation of ΔP_T along the module length remained a formidable obstacle until the recent development of the Bactocatch process by Alfa-Laval (Sandblom 1978) and its application to milk filtration (Malmberg and Holm 1988, Olesen and Jensen 1989).

The principle and characteristics of microfiltration with uniform trans-membrane pressure, also known as the Bactocatch process, are described in Chapter 5 (see Section 5.6). The application of this technique to cross-flow filtration using inorganic membranes for the removal of bacteria from skimmed milk and protein permeation (or recovery) are discussed next, along with a comparison of filtration performance using conventional cross-flow filtration.

The application of the Bactocatch® process for the production of 12,000 L/h of low-fat (0.5 wt.%) and medium-fat (1 wt.%) milk was described (Malmberg and Holm 1988). An average flux of about 500 L/h-m², at a concentration factor of 10, was reported with a bacteria removal efficiency of greater than 99.6%. For low-fat skimmed milk, flux values as high as 750 L/h-m² at a concentration factor of 20 can be obtained (Maubois 1990). The regeneration of fouled membranes was done after 5–10 h of continuous operation.

Membrane cleaning is also easier when filtration is performed under relatively constant ΔP_T values as compared to membranes fouled under conventional cross-flow filtration configuration. The optimal increase in ΔP_T values to maintain the desired flux and operation time were found to lie in the range 0.05–0.4 bar. The optimal ΔP_T values for a 1.4 μm alumina membrane (Membralox® 1P19-40 multichannels, Alcoa/SCT) were found to be in the range 0.3–0.8 bar (Olesen and Jensen 1989).

The influence of operation parameters on bacteria retention and the physicochemical properties of the processed milk using the Bactocatch process was also studied (Olesen and Jensen 1989). These authors investig-ated the performance using a 2-level factorial design of experiments by varying the parameter values within a specified range consistent with values used in industrial practice. Experimental results showed that the spore count of *Bacillus cereus* in the initial milk had a significant effect on the spore count in the microfiltered milk.

The variations in the concentration factor, ΔP_T, treatment temperature for retentate (containing concentrate along with bacteria) and pasteurization did not show any significant effect on the bacteria content of the microfiltered milk. A reduction of 99.99% in the total bacteria count along with greater than 99.95% reduction in the spore count was achieved. Although a few bacteria may pass into the permeate, their total count was always found to be within the acceptable limits. Alfa-Laval, the developer of the Bactocatch process, guarantees, a bacteria removal of at least 99.6%. The rennetability (casein content) of microfiltered milk was reduced by less than 10% with a minor increase in the concentration of casein in the retentate. These had virtually no impact on the usability of the microfiltered milk.

On the other hand, rapid fouling and lower flux resulted when the milk was filtered under conventional cross-flow filtration, where the permeate-side pressure was held constant as the feed-side pressure decreased from inlet to outlet. This is illustrated in Figure 8.2. It is evident that an optimal ΔP_T value can be maintained only over a small fraction of the filtration area (estimated to be about 20%) as compared to almost 100% for the uniform ΔP_T for the Bactocatch process.

In addition to the higher flux characteristics, there are other practical benefits/advantages for low-bacteria milk produced by the Alfa-Laval Bactocatch process. An almost 100% reduction in the total bacteria count and particularly the spore count of *Bacillus cereus* increases the shelf-life of Bactocatch-treated milk from a typical refrigerated shelf-life of 6–8 days to a higher range of 16–21 days (Malmberg and Holm 1988). Longer shelf-life

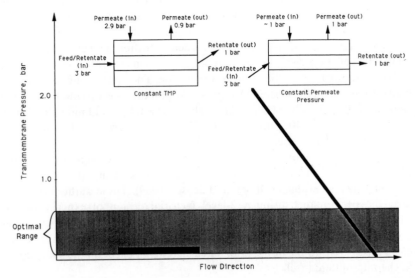

Figure 8.2. Typical pressure profiles for filtration under constant transmembrane pressure (TMP) and constant permeate pressure (Malmberg and Holm 1988).

means greater safety margins for the producer and increased convenience for the consumer.

8.2.2. Concentration of Pasteurized Skimmed Milk

One of the most promising applications where inorganic membranes have enjoyed significant success is the concentration of pasteurized skimmed milk also known as milk protein standardization. For this application, a 0.2 μm membrane was typically used. Membralox® alumina membranes and Carbosep® membranes consisting of a ZrO_2 layer on carbon support are used commercially. Figure 8.3 shows a typical industrial plant for the production of pasteurized skimmed milk. In this process, skimmed milk is pasteurized at about 70°C and concentrated twofold. Higher concentration factors up to 5 have also been achieved using a 0.2 μm alumina membrane with excellent results (Attia, Bennasar and Tarodo 1988).

Recently, 0.1 μm Membralox® zirconia membranes were also evaluated for this application (Tarodo and Lecornu 1988). The cross-flow velocity showed a significant impact on flux. The range of cross-flow velocity normally recommended in this application is 5–6 m/s. Figure 8.4 shows the effect of cross-flow velocity on flux for the 0.1 μm zirconia membrane. At a cross-flow velocity of 5.5 m/s, a stable flux of about 100 L/h-m^2 was obtained. The operating temperature was 50°C and the value of ΔP_T was 5 bar. This flux value is higher by a factor of 1.5–2 as compared to that typically obtained with a 0.2 μm pore diameter membrane (e.g. Membralox® or Carbosep®). Several industrial plants in operation since early 1989 confirm the performance data on Membralox® ceramic membranes described above obtained in pilot-scale process evaluations.

The concentration of pasteurized skimmed milk may also be performed for phosphocaseinate separations. The permeate from such a process can serve as an ideal whey without fat, bacteria, rennet and glycomacropeptide (Maubois 1987).

8.2.3. Concentration of Whole Milk

The concentration of raw whole milk and/or pasteurized whole milk using ceramic membranes is practiced on the commercial scale using membranes with a pore diameter of 0.1 and 0.2 μm (Bennasar and Tarodo 1983). The ceramic membranes are operated in the cross-flow configuration. The primary purpose of this process is to produce a liquid precheese subsequently used in the manufacture of soft or semi-hard cheese.

The unique advantage of inorganic membranes lies in the fact that only these are capable of producing a precheese product with a protein content of

Figure 8.3. A typical milk protein standardization plant using Membralox® ceramic membrane filters of approximately 50 m² membrane area (courtesy of SCT, Tarbes, France).

21% or higher (Bennasar et al. 1984a, Gillot and Garcera 1986, Goudedranche et al. 1980).

It is known that ultrafiltration of whole milk at low temperatures, using alumina or zirconia membranes with pore diameters in the range 18 nm to 0.2 μm, is limited by concentration polarization and/or gel polarization. The polarization layer consists casein micelles, whey proteins, salts and fat and

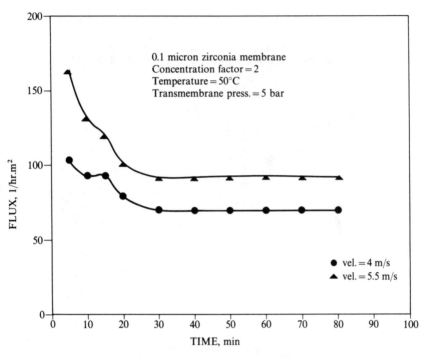

Figure 8.4. Performance of Membralox® ceramic membranes in the concentration of pasteurized skimmed milk (Tarodo and Lecornu 1988).

acts as a prefilter to the membrane-separating layer. UF flux can be increased by reducing the boundary layer (Vetier, Bennasar and Tarodo 1988, Bennasar and Tarodo 1983, see also Section 4.3.2). In addition, there are several factors/parameters that may have a significant effect on flux, such as flow regime (laminar versus turbulent or transitional), prerennetting (fat enrichment), enrichment of milk with calcium and thermal pretreatment.

The retention of proteins is excellent and comparable to polymeric membranes thus demonstrating that concentration of milk with inorganic membranes is economically and technologically viable. Also, the nutritional value is comparable to that obtained using other membrane materials or process technologies.

Using a 0.2 μm Membralox® at an operating temperature of 5°C, permeate flux of 50 L/h-m² was obtained with a volumetric concentration factor of 2 (Bennasar et al. 1984b). This value is about 2.5 times higher than that reported using polymeric membranes (Slack et al. 1982). The protein retention was about 93%. This is based on total nitrogen analysis. The actual protein retention was about 98%. This is due to the fact that about 5% is non protein nitrogen.

Recently, a 0.1 μm Membralox® zirconia membrane was evaluated for the concentration of raw whole milk and pasteurized whole milk (Tarodo and Lecornu 1988). The results are shown in Figure 8.5. It is evident that the zirconia membrane performance is far superior to that of the 0.2 μm alumina membrane described earlier. For raw whole milk, at a concentration factor of a stable flux of about 81 L/h-m² was sustained and about 78 L/h-m² for the concentration of pasteurized whole milk.

Membrane regeneration with an alkaline detergent solution (containing NaOH) followed by a sanitizing acid solution (1–2% HNO₃) was very efficient and ensured that the original water flux was restored after cleaning. Polymeric membranes are relatively difficult to clean due to their limited chemical resistance. The additional use of chlorine and chemical complexants improve cleaning efficiency.

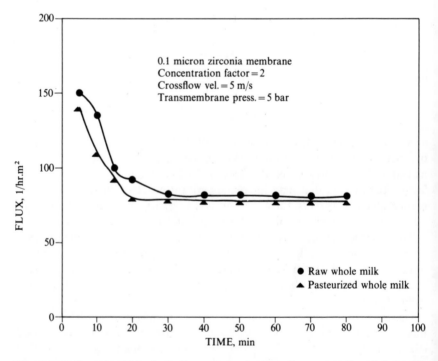

Figure 8.5. Performance of Membralox® ceramic membranes in the concentration of whole milk (Tarodo and Lecornu 1988).

8.2.4. Microfiltration in the Processing of Whey to Produce Whey Protein Concentrate

In recent years, microfiltration (MF) of whey is being considered for the removal of bacteria, lipids (viz. lipoproteins and phospholipids) and casein fines (Maubois et al. 1987, van der Horst 1990). Microfiltration of whey is a relatively new approach that has not yet found commercial applications. However, several inorganic and polymeric membranes are being evaluated on the pilot scale (Merin and Daufin 1990). Current technology/material alternatives to the use of inorganic membranes for the concentration of whey include reverse osmosis (RO), evaporation or use of polymeric UF membranes.

Microfiltered whey may be used as the feed in the subsequent ultrafiltration concentration step in the manufacture of whey protein concentrate (WPC) powder to increase the percentage of proteins/total solids to about 50.

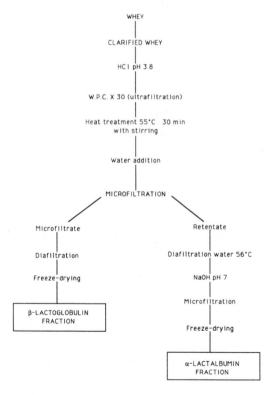

Figure 8.6. Separation of α-lactalbumin and β-lactoglobulin from defatted whey protein concentrate (Maubois et al. 1987).

This is typically designated as WPC 50. It is also thought that such a lipid-free concentrate may serve as an excellent feed for further purification and separation into individual whey proteins (van der Horst 1990). The microfiltration process in combination with diafiltration can also be utilized for the separation of α-lactalbumin and β-lactoglobulin from defatted WPC (Maubois et al. 1987). A schematic for the separation of these two proteins is shown in Figure 8.6.

Microfiltering the whey was shown to improve flux for the subsequent UF concentration step. In a recent study reported by van der Horst (1990), flux during MF of concentrated whey was found to be up to 200% higher than the flux value obtained during UF of concentrated whey.

The comparison of ceramic versus polymeric membrane performance for cross-flow filtration showed that ceramic membranes gave reliable reproducible results combined with higher flux and separation characteristics. The ceramic membranes evaluated were 0.2 μm Membralox® (Alcoa/SCT) alumina membrane and 0.14 μm pore diameter Carbosep® (Tech Sep) zirconia MF membranes. The polymeric membranes used in the study were 0.1 μm Abcor (Koch) polyether sulfone membranes and 0.1 μm pore diameter Amicon (W.R. Grace) polysulfone membranes. The results are summarized in Table 8.1.

Table 8.1. Performance of Ceramic and Polymeric Membranes in the Microfiltration of Whey Concentrated by Reverse Osmosis (van der Horst 1990)

Membrane	Membrane/ Support Material	Nominal Pore Size (μm)	Transmembrane Pressure (bar)	Cross-flow Velocity (m/s)	Flux (L/h-m²)	Protein Permeation (%)	Fat/ Protein (mg/g)
Amicon	Polysulfone	0.1	1.0	1.3	25	18	3.3
Alcoa/SCT	Al₂O₃	0.2	1.1	2.5	28	12	1.5
			1.1	5	53	34	1.4
			2.1	5	64	27	3.4
Abcor	Polyether sulfone	0.1	1.0	0.7	36	40	1.2
			2.3	2.3	65	34	1.1
			3.0	0.7	40	17	3.6
Tech Sep	ZrO₂/C	0.14	1.0	2.5	17	33	1.0
			1.0	5.0	19	14	2.3
			1.2	7.5	36	42	1.9
			2.5	5.0	59	14	2.4
			3.2	7.5	69	15	2.8

25% total solids (wt./wt.)

The average flux for RO concentrated whey increased by 80% with an increase in protein retention of about 46% compared to whey concentrated by evaporation. On the other hand, MF of whey did not improve flux or protein retention. The total solids content of the feed was about 25 wt.%. The addition of $CaCl_2$ to improve the size of the aggregates also proved ineffective in improving flux and often caused severe membrane fouling. It was concluded that conventional heat treatment at 55°C was a better option to improve flux and protein recovery using RO membranes.

Under optimal processing conditions (cross-flow velocity = 5 m/s, $\Delta P_T \sim 2$ bar, 50°C), the highest flux with 0.2 μm Membralox® alumina membranes was 64 L/h-m², whereas for Carbosep® membranes it was 59 L/h-m². Protein permeation with alumina membranes was 27% as compared to 14% for the zirconia membranes. Flux and protein recovery values for Koch membranes under optimal conditions (cross-flow velocity = 2.3 m/s, ΔP_T = 2.3 bar) were 65 L/h-m² and 34%, respectively. Amicon membranes gave the lowest flux and separation performance. At a cross-flow

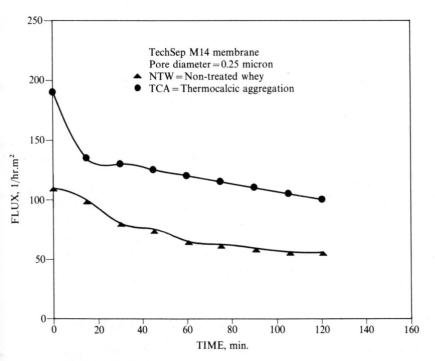

Figure 8.7. Effect of thermocalcic aggregation on flux in the separation of whey proteins (Maubois et al. 1987).

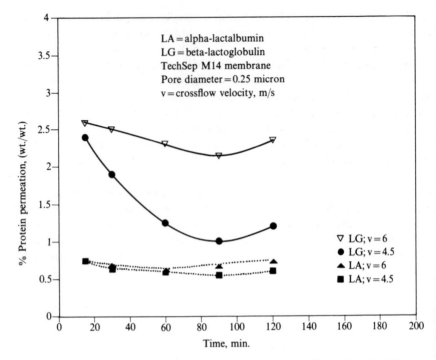

Figure 8.8. Effect of cross-flow velocity on protein permeation (Maubois et al. 1987).

velocity of 1.3 m/s and a ΔP_T of 1 bar, permeate flux was only about 25 L/h-m^2 with a protein permeation of 18%.

In a separate study, Maubois et al. (1987) reported the development of a pretreatment step involving thermocalcic aggregation which substantially improved the flux during MF. A flux increase by about 80% to a value of 110 L/h-m^2 was achieved as compared to the flux of about 60 L/h-m^2 obtained with nontreated whey. The results are shown in Figure 8.7. The effect of cross-flow velocity on protein permeation is shown in Figure 8.8. This work clearly shows the importance of cross-flow velocity and pretreatment in the processing of whey to produce WPC, provided the operating conditions are compatible with the stability of molecules.

8.2.5. Concentration of Serum Proteins from Whey by Ultrafiltration

As in the case of microfiltration of whey discussed in Section 8.2.4, the use of inorganic UF membranes on the commercial scale for serum proteins concentration is still under development (Merin and Daufin 1990, Daufin et al. 1990, Daufin, Michel and Merin 1990). Several industrial-scale optimization studies are reported (Maubois et al. 1987, Daufin et al. 1990, Daufin, Michel and Merin 1990).

Membrane-based ultrafiltration has been used commercially in the food industry for over two decades. UF has been very successfully applied to concentrate whey proteins to obtain WPC fractions (in powder form) with different levels of purity ranging from 35% up to 95%. Alumina-based and zirconia-based UF membranes with pore diameters in the range 20–100 nm (Membralox®, Alcoa/SCT) as well as ZrO_2 on carbon membranes with a molecular weight (MW) cutoff of 10,000 (approximate pore diameter 4–5 nm) appear to be suitable for whey protein concentration.

In a recent study, the concentration of whey proteins from sweet whey and sour (acidified) whey using 100 nm Membralox® zirconia UF membranes was described (Tarodo and Lecornu 1988). The results are reported in Figure 8.9. At a volumetric concentration factor of 20, flux values in the range 50–60 L/h-m² were obtained. This corresponds to a protein concentration of about 25% (wt./wt.). The ultrafiltration of acid whey to obtain a 37 wt.%

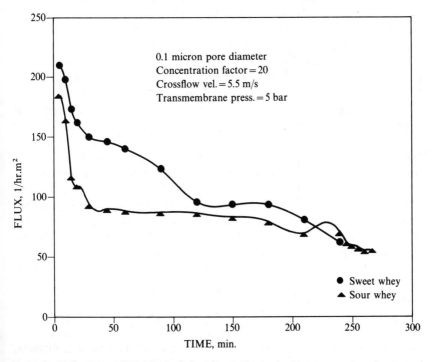

Figure 8.9. Performance of Membralox® zirconia membranes in the concentration of sweet and sour whey (Tarodo and Lecornu 1988).

protein concentrate using 10,000 MW cutoff ZrO_2 on carbon membranes (Carbosep® M5) manufactured by Tech Sep was also reported (Merin and Daufin 1990). In another investigation, UF was used in conjunction with diafiltration to obtain WPC of high purity in the range 70–95% (Maubois et al. 1987). The whey was pretreated by MF prior to UF using the technique of thermocalcic aggregation where larger sized particles (casein) are formed to ease filtration and flux enhancement. Taddei et al. (1989) showed that concentration of β-lactoglobulin, one of the main proteins in milk whey, could not be raised beyond 60% with a 20,000 MW cutoff zirconia membrane (Carbosep® M4). However, with the use of a 10,000 MW cutoff zirconia membrane (Carbosep® M5), high protein concentrations up to 85–95% could be achieved (Daufin et al. 1990). These results were obtained with almost 100% protein retention.

8.2.6. Concentration of Acidified Milk to Produce Fresh Cream Cheese

The production of cream cheese by the chemical acidification of skimmed or whole skimmed milk is practiced on the commercial scale (Mahaut et al. 1986). The process involves the separation of an aqueous phase containing lactose and minerals as the filtrate from a semisolid phase where all the proteins and fat are located in the product phase (retentate). Conventionally, centrifugal separators are widely used to concentrate acidified milk to produce cream cheese. Cream cheese is one of the many cheese varieties (e.g. Pate fraiche, Petit Suisse, Quarg, St. Paulin etc.) now amenable for production using inorganic membranes (Mahaut et al. 1982, Goudedranche et al. 1980).

A major advantage in using a stand-alone inorganic ultrafiltration membrane system for the production of cream cheese is its potential to replace the traditional process where organic UF membranes are used in combination with either classical evaporation or scraped surface evaporation (Maubois, Mocquot and Vassal 1974, Goudedranche et al. 1980). This advantage, however, must be compared and balanced with the relatively larger energy requirements associated with inorganic tubular UF systems having channel diameters ranging from 3–6 mm.

Acidification is necessary to solubilize calcium salts and ensure that the calcium content and organoleptic qualities are similar to those found in traditionally prepared cheese. Typical protein retention for chemically acidified lactic curds, with inorganic membranes such as 0.2 μm Membralox® alumina membranes, is about 95–96% with a flux of 120 L/h-m^2 at 21 wt.% protein concentration (Rios et al. 1989). Membralox® 100 nm pore diameter zirconia membranes gave a flux of 95 L/h-m^2 at 50°C (volumetric concentration factor = 2) with greater than 98% protein retention (Alcoa/SCT 1990).

In a recent study, a 100 nm Membralox® zirconia membrane was evaluated for the concentration of acidified (curdled) milk (Tarodo and Lecornu 1988). At a volumetric concentration factor (VCR) of 2, flux values averaged about 145 L/h-m² with 98.4% protein retention. The results are shown in Figure 8.10. At the higher VCR of 4, flux values for the 0.1 μm zirconia membrane averaged significantly lower and were below those obtained using 0.2 μm alumina membranes. These data are reported in Figure 8.11.

Another requirement in the production of semi-hard cheese is the high concentration of proteins (approximately 20–22%) in the precheese prior to UF. This is difficult to achieve with polymeric UF membranes due to mechanical compaction of the membrane and the high pressure drop that results in a very low flux (Korolczuk, Maubois and Fauquant1987, Goudedranche et al. 1980). Additionally, polymeric membranes pose difficulties in cleaning.

Analogous to many milk processing applications such as raw whole milk and reconstituted skimmed milk (Vetier, Bennasar and Tarodo, 1986, 1988), the flux in the filtration and concentration of acidified milk is limited by the

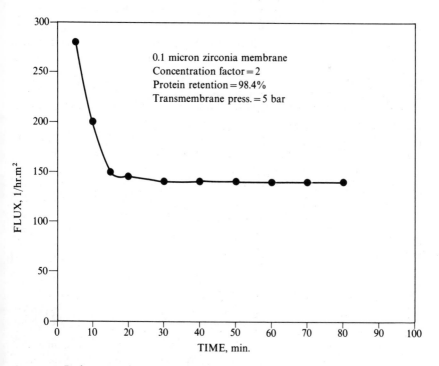

Figure 8.10. Performance of Membralox® ceramic membranes in the concentration of acidified (curdled) milk (Tarodo and Lecornu 1988).

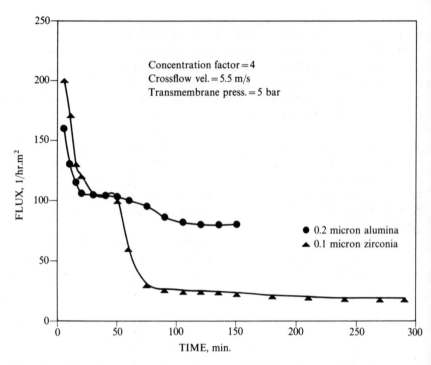

Figure 8.11. Effect of pore size on the performance of Membralox® ceramic membranes in the concentration of acidified (curdled) milk (Tarodo and Lecornu 1988).

formation of porous deposit on the membrane surface from the interactions between the membrane and milk components. (Attia, Bennasar and Tarodo de la Fuente 1991a, b). The fouling layer is characterized as consisting of casein soluble proteins such as α-lactalbumin, β-lactoglobulin, bovine serum albumin, calcium and phosphorous salts. Some of these aspects are further discussed in Section 4.3.2.

The effect of hydraulic parameters on filtration performance was investigated by some research groups in France (Goudedranche et al. 1980, Rios et al. 1989, Vetier, Bennasar and Tarodo 1986). Studies with Membralox® 0.2 and 0.8 μm pore diameter Al_2O_3 membranes showed that fouling was less pronounced with 0.2 μm membranes as compared to 0.8 μm membranes. This result can be expected as the relatively large protein molecules (up to 0.3 μm; Muir and Banks 1985) can cause severe pore blockage of 0.8 μm membranes as compared to the smaller 0.2 μm membranes. It was also observed that in order to prevent excessive pressure drop (> 6–8 bar), it is beneficial to use a larger channel diameter (e.g. 6 mm) in place of the commonly used multichannel elements with 4 mm channel diameter.

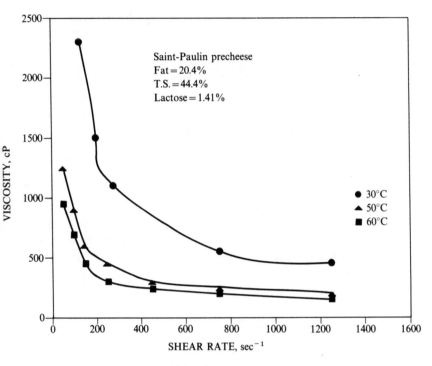

Figure 8.12. Variation of viscosity with shear rate (Goudedranche et al. 1980).

Retentate viscosity is another parameter of great significance. Viscosity increases (to as high as 200 mPa-s) as the retentate is progressively concentrated. It can, however, be reduced by increasing the shear rate (i.e. cross-flow velocity in cross-flow filtration) and temperature. A typical relationship between viscosity and shear rate, at various operating temperatures, observed in the production of St. Paulin precheese is shown in Figure 8.12 (Goudedranche et al. 1980). Inorganic membrane structures are uniquely suited to handle the mechanical stress generated at high retentate viscosities. Although flux can be increased by increasing the temperature, protein retention is adversely affected. Operating temperatures above 50°C were also found to reduce the organoleptic qualities of the product to below acceptable levels (Rios et al. 1989).

8.2.7. Rheological Behavior of Concentrates During the Processing of Lactic Curds Using Inorganic Membranes

Coagulums with an initial dry matter (DM) content of 9.7% (wt./wt.) obtained by (1) lactic fermentation only [starter concentration 5×10^{-3}%

(wt./vol.), 25°C], (2) lactic fermentation and rennet addition [starter concentration $5 \times 10^{-3}\%$ (wt./vol.), rennet at 520 mg/L of chymosin = 0.25 $\times 10^{-3}\%$ (vol./vol.), 25°C] or (3) direct acidification (0.2 N HCl, pH = 4.4, 20°C) were concentrated on inorganic membranes (average pore diameter of 0.2 μm) to a final dry matter content of 20.5% (wt./wt.). The variations in permeate flux, retentate pressure drops (ΔP) and protein retention as a function of retentate dry matter or concentration factor, for each of the above cases, appeared to be noticeably different (Tarodo 1990a).

Flux values for fermented coagulums, with or without rennet addition, declined quickly when the DM content increased before its stabilization, whereas flux values for coagulums obtained by direct acidification remained almost constant except at the higher DM values where they showed a nominal increase. These results are shown in Figure 8.13 (Tarodo 1990a). The stabilization of flux at the end of the process is probably attributable to the considerable simultaneous increase in pressure drops which enhance solvent transfer. These head losses were greater with biological coagulum because of the relatively higher viscosity.

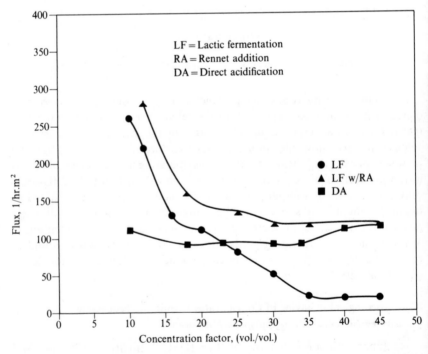

Figure 8.13. Concentration of lactic curds: flux versus concentration factor (Tarodo 1990a).

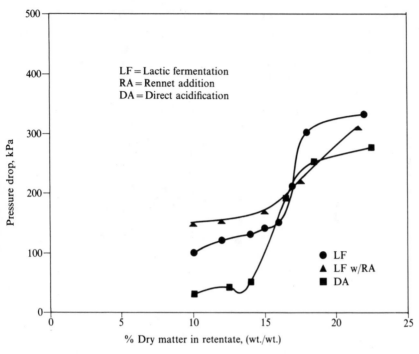

Figure 8.14. Concentration of lactic curds: pressure drop versus dry matter content (Tarodo 1990a).

Pressure drops were similar in all cases and displayed a sigmoidal shape with a point of inflection which corresponded approximately to the end of the fall in permeate flux. These data are plotted in Figure 8.14 (Tarodo 1990a).

The variation of protein retention with the percentage of dry matter in retentate is shown in Figure 8.15 (Tarodo 1990a). Protein retention values were not very significantly different. They showed a tendency to increase for DM values up to about 15% and decreased somewhat at higher values of retentate dry matter content. Protein retention values were highest for coagulums obtained by lactic fermentation.

The differences in performances observed between the coagulums studied were undoubtedly due to differences in fouling which in turn resulted from variations in the preparation and structure of the initial products. The low initial flux and early attainment of a "state of equilibrium" in the case of chemical coagulum may be caused by rapid saturation of adsorption sites at the surface of alumina particles by proteins.

Protein retention has several causes: (1) the mean transmembrane pressure (ΔP_T) which makes the deposit settle, reduces its porosity and limits the

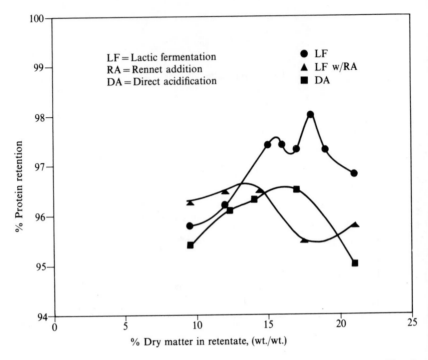

Figure 8.15. Concentration of lactic curds: protein retention versus dry matter content (Tarodo 1990a).

passage of large proteins, (2) the osmotic pressure which increases with concentration and enhances the passage of small protein molecules and (3) the possibility of interactions in external fouling. The latter effect may account for the relatively high protein retention observed in renneted milk coagulum.

Rheograms [apparent viscosity (η) in relation to shear rate (D) at 50°C] of the initial products and retentates as a function of their concentration (%DM content) are shown for fermented coagulums of renneted milk in Figure 8.16 (Tarodo 1990a). The initial coagulum displayed quasi-Newtonian behavior since its viscosity was almost independent of shear rate. Subsequently, it displayed non-Newtonian behavior which was all the more marked when the DM content of the retentate was higher. Other coagulums showed similar behavior. With a shear rate constant of $662\,\mathrm{s}^{-1}$, viscosities of all retentates increased exponentially with DM content and reached 92×10^{-3} Pa-s for direct acidified coagulums, 205×10^{-3} Pa-s for fermentated coagulums and 254×10^{-3} Pa-s for fermented coagulums of renneted milk. All retentates

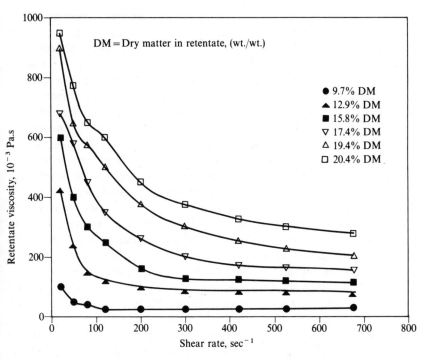

Figure 8.16. Concentration of lactic curds: retentate viscosity versus shear rate (Tarodo 1990a).

displayed a threshold stress which also increased exponentially with the DM content.

Whatever their origin, retentates in the process of concentrating thus displayed the behavior of plastic bodies since they displayed a threshold stress which is a qualitative property of these bodies. Liquids only flow beyond a certain shear stress. At lower values, these retentates behave like solids. This threshold stress probably corresponds to the force required to cause a certain separation of protein particles (which are bound together or have become interlocked during flocculation) and to overcome cohesive forces and thus cause flow. In addition, this non-Newtonian behavior is all the more marked when the concentration increases. In this situation, the biological retentates (with or without rennet) were 2–2.5 times as viscous as those obtained by the direct acidification route, at the same DM value.

This proves that the aggregation of protein particles is different in the two types of coagulums (biological and chemical). This can be accounted for by invoking a smaller number of weaker bonds in the chemical retentate. These bonds probably offer less resistance to the orientation of proteins in the direction of flow. This hypothesis is confirmed by the lower shear stresses at the threshold stress in the case of chemical coagulum, and supported by the study of evolution of viscosity as a function of time for final retentates obtained with the various coagulated products. The higher viscosities of the retentates obtained exclusively by the biological coagulation of renneted milk as compared to those obtained without biological coagulation may result from the presence of bonds between sites released by the coagulant enzyme. This result tends to agree with the fall in performance and the increase in efficiency observed during the concentration of renneted milk coagulum.

It is necessary to know and take into account these rheological features in the design of ultrafilters. Thus, in the case of a multistage process where high concentrations are realized, a process interruption or failure will cause rapid changes in the retentate properties, such as from fluid to paste with all the characteristics of a gel with a high threshold stress that is very difficult to remove. This phenomenon was observed by Goudedranche et al. (1980), who showed that the accidental failure of circulation in the finishing stage resulted in a considerable increase in the viscosity of St. Paulin cheese retentate. Mahaut et al. (1982) made the same observation in the concentration of lactic coagulum. In addition, it is important to be able to determine the desired shear rate at each stage of an industrial installation in relation to concentration to obtain a retentate viscosity making it possible to operate in the desired flow regime. In the present case, the very high viscosities of the retentates systematically cause laminar flow from the beginning of the concentration process (the Reynolds number can drop from 1000 to 50 during concentration).

The characteristic rheograms [shear stress (τ) as a function of shear rate (D)] of the final retentates of the various coagulums are shown in Figure 8.17 (Tarodo 1990a). The most representative rheological equations for the fermentated coagulums are those which combine a power law ($\tau = KD^n$) and a threshold stress (τ_s). This is Herschell–Bukley's law.

$$\tau = \tau_s + KD^n \quad \text{where } n < 1 \tag{8.1}$$

K is the consistency index and n is the behavior index.

In the case of chemical coagulums, the rheogram of the final retentate revealed two different flow regimes as illustrated in Figure 8.17. At shear rates of less than $300\,\text{s}^{-1}$, the flow was Newtonian. At approximately $300\,\text{s}^{-1}$, the viscosity changed sharply and the flow became non-Newtonian at shear rates greater than $300\,\text{s}^{-1}$.

The evolution of apparent retentate viscosity as a function of time is shown in Figure 8.18 (Tarodo 1990a). At a shear rate of $662\,\text{s}^{-1}$, there is a considerable initial fall in viscosity which occurs for a period of approximately 30 min. This is then followed by a less rapid decrease in coagulum of

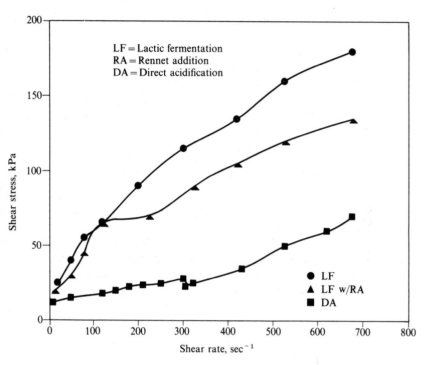

Figure 8.17. Concentration of lactic curds: shear stress versus shear rate (Tarodo 1990a).

renneted milk, a tendency towards stabilization for fermentated coagulum and a plateau for chemical coagulum.

These results confirm the differences in the nature of the bonds responsible for the aggregation of proteins and the structure of coagula according to type of coagulation. They are weak in chemical coagulum, medium in biological coagulum and strong in coagulum of renneted milk. Orientation of protein particles in the direction of flow is thus easier and more rapid for chemical coagulum. Beyond a period of 30 min, its microscopic state does not change any more, as is proved by the steadiness of apparent viscosity in time. This coagulum is, therefore, not thixotropic (Grossiord and Couarraze 1983). However, biological coagulums display structural instability characteristic of thixotropic behavior of these coagulums. Nevertheless, before making definitive statements on the possible thixotropic behavior of these coagula, it should be checked if their initial structures recover fully after a rest period of a few minutes to several days.

It is important to know the rheological characteristics of the retentates for the choice of the texture of the final product. Indeed, the latter is determined

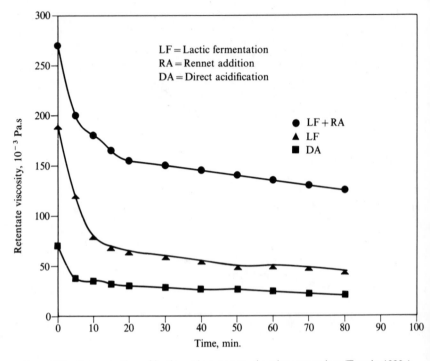

Figure 8.18. Concentration of lactic curds: retentate viscosity versus time (Tarodo 1990a).

by two imposed factors, one dynamic (shear stress) and the other kinematic (shear rate). From these factors stem the plastic properties such as the threshold stress, and thixotropic properties such as low viscosity under stress and return to the initial structure. Consequences of these properties can be illustrated by: a paste which does not flow when it is at rest, which spreads under a stress (knife) without leaving marks and which recovers to its initial solid appearance. The rapid, important variations in viscosity during the first few minutes of subjection to shear should also be taken into account in the definition of the conditions for starting up an ultrafiltration installation. These results can be accounted for by the different properties of the initial products (depending on the type of coagulation) which cause differences in the formation and structure of fouling.

Observations made with the help of scanning electron microscopy for membrane samples coming into contact with the various coagula revealed rapid massive adsorption of fine fouling particles on the alumina grains of the filter layer in the case of chemical coagulum. On the other hand, for biological coagulums adsorption was reduced to a few large, well-separated aggregates.

The analysis of rheograms made it possible to monitor the influence of the acidification method on the viscosity and flow threshold of the retentates. With the same dry matter content, the viscosities of the retentates of biological coagulums (with or without rennet) were distinctly higher than those of the chemical coagulum. The flow thresholds were also greater for biological coagulums than for the chemical coagulums. This can be explained by the forces of cohesion between the particles in biological coagulums which are larger than those observed for chemical coagulums.

The determination of rheological equations for the final retentate (DM = 20.5% wt./wt.) and evolution of viscosity in relation to time showed that biological coagulums (with or without rennet) behave as nonideal plastic bodies (Herschell–Bukley law) whereas a chemical coagulum behaves as an ideal plastic body (Bingham body) or a Newtonian body, depending on whether the shear rate is less than or greater than $300 s^{-1}$. These various rheograms and rheological equations can make valuable contributions from a practical point of view, by making it possible (1) to size the various parts of the ultrafiltration installation (pumps, piping, heat exchanger, stirring device, etc.) according to the rheological characteristics (high viscosity, threshold stress) of the coagulums and their hydrodynamic consequences (laminar flow), (2) to obtain knowledge of and, possibly, control the properties of the finished product (appearance, texture, viscosity, spread potential, etc.) in relation to operating conditions.

8.3. INORGANIC MEMBRANES IN THE CLARIFICATION OF FRUIT JUICES

Traditionally, polymeric membranes have been used in the clarification of various fruit juices such as apple juice, cranberry juice, grapefruit juice and kiwifruit juice. However, these processes suffer from some loss of juice flavor during clarification.

The technological advantages offered by inorganic membranes, such as ceramic and inorganic membrane layers on metallic supports, are higher flux, minimal (or nil) protein adsorption and absolute cutoff characteristics. Inorganic filters can also be steam sterilized in-line, backflushed at high pressures, and they feature unlimited autoclavability. Inorganic filters are chemically inert and, unlike their polymeric counterparts, will not compact, swell or creep-flow. These properties, in addition to their high strength, temperature resistance and lack of media migration, separate them from the traditionally used polymeric membranes in many applications. Typical operating parameter values for the clarification or concentration of various fruit juices are given in Table 8.2 (Venkataraman, Silverberg and Giles 1988).

Table 8.2. Typical Operating Parameters Used in the Clarification of Various Fruit Juices (Venkataraman, Silverberg and Giles 1988)

Juice	Flux (L/h-m^2)	Membrane Pore size (microns)	Operating Temperature (°C)	Transmembrane Pressure (bar)	Cross-flow Velocity (m/s)
Apple juice concentrate (70–71 Brix)	17.5	0.2	70–75	5	4.9
Apple juice (1–2% solids; 1% Pectin)	250	0.2	22	3.5	4.9
Apple juice (12.5 Brix; 1% insolubles; 1% pectin)	460	0.2	75	4	6.7
Orange juice (10.5 Brix)	50	0.2	15	1.5	4.4
Cranberry juice clarification (5% solids)	167	0.45	50	1.5	4.9
Red fruit juice (12–12.4 Brix)	20	0.45	50	3.5	4.6
Pineapple (mill) juice clarification (7% solubles)	240	0.20	50	7	5.3
Pink grapefruit juice	22	0.20	25	1.5	4.9

All % values in (wt./wt.)

8.3.1. Apple Juice Clarification Using Ceramic Membranes

The use of ceramic membranes in the clarification of apple juice is one of the most successful and widely practiced industrial applications. The longer operating life span for ceramic membranes together with high-quality filtered juice, obtainable at filtration rates that are comparable to and often exceeding those of competing products/technologies such as diatomaceous earth filters and synthetic polymeric membranes, have been very central to their success.

Industrial plants with as large as 160 m^2 membrane areas have been in continuous operation for over 4 years without membrane replacement. The ceramic membrane modules are typically installed in several loops operating in parallel. The overall process may consist of several stages depending on the level of juice concentration desired. Figure 8.19 shows an industrial installation for the production of clarified apple juice.

Pore diameters of the composite ceramic membranes used in apple juice clarification range from 0.2–0.8 μm. A number of commercial membranes are

Figure 8.19. A typical industrial-scale apple juice clarification plant using Membralox® ceramic membrane filters (70 m²). (courtesy of MEMBRAFLOW Filtersysteme, Germany).

available for this application (Membralox®, Ceraflo™ and Carbosep®). The average flux with a 0.2 μm membrane (e.g. Membralox® alumina membrane) for non-preconcentrated juice is in the range 180–200 L/h-m². For a twofold preconcentrated juice, average flux values in the range 100–150 L/h-m² are realized (Guibaud 1990).

Recently, 0.1 μm Membralox® zirconia membranes were evaluated for the concentration of apple must. Flux values as high as 250 L/h-m² were obtained at a concentration factor of 10 (Tarodo and Lecornu 1988). The results are reported in Figure 8.20. The filtrate quality was excellent. These results clearly show the superior performance of 0.1 μm Membralox® zirconia membranes as compared to 0.2 μm membranes currently used in industry. In all cases, enzymatic pretreatment (pectinases) has a positive effect on the permeate flux.

8.3.2. Clarification of Apple Juice Using Inorganic Membranes on Porous Metallic Supports

The conventional process for producing apple juice requires separate operations for processing, filtering and pasteurization. Pasteurization temperatures and filtration of the juice degrade the juice flavor.

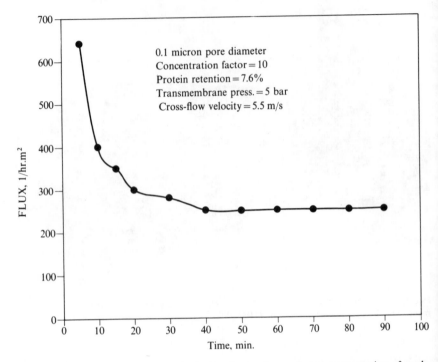

0.1 micron pore diameter
Concentration factor = 10
Protein retention = 7.6%
Transmembrane press. = 5 bar
Cross-flow velocity = 5.5 m/s

Figure 8.20. Performance of Membralox® zirconia membranes in the concentration of apple must (Tarodo and Lecornu 1988).

Carre Inc. (a subsidiary of Du Pont Separations) and the Food Science Department of Clemsen University developed and patented a process that uses an inorganic membrane on a microporous stainless steel support (Carre 1986). The Ultrapres™ membrane process combines the pressing and juice clarification in a single-unit operation. The produced juice is sterile with a flavor and aroma superior to that produced by other processes, including those using polymeric membranes. The production of apple juice in a single stage using inorganic membranes on metallic supports was also recently reported (Thomas 1988).

The Ultrapres™ process operates with either a feed of whole apples pulverized into small pieces or a "finished" mash where seeds, stem and peels are separated from the pulp. The schematic of a laboratory-scale Ultrapres™ process is shown in Figure 8.21. This process can also be used for the production of other fruit juices including pineapple, pear, peach and kiwifruit (Carre 1986). Typical yields of about 85% are obtained in a single-stage process (Thomas et al. 1986). The typical ΔP_T values used are in the

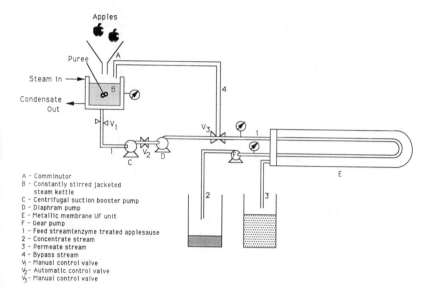

Apples

Puree

A

Steam in

Condensate
Out

B

V₁

V₃

1

4

C

V₂

D

E

A – Comminutor
B – Constantly stirred jacketed
    steam kettle
C – Centrifugal suction booster pump
D – Diaphram pump
E – Metallic membrane UF unit
F – Gear pump
1 – Feed stream(enzyme treated applesause)
2 – Concentrate stream
3 – Permeate stream
4 – Bypass stream
V₁ – Manual control valve
V₂ – Automatic control valve
V₃ – Manual control valve

Figure 8.21. Metallic membrane ultrafiltration system for apple juice clarification (Thomas et al. 1986).

range 15–20 bar which yield flux values in the range 50–70 L/h-m^2. The channel diameter of the membrane tube is about 0.03 m (Thomas 1988). Membranes formed on porous metallic supports can be steam sterilized and can tolerate a wide pH range of 2–13.

The permeate flux of a hollow-fiber or thin-channel polymeric membrane declines appreciably during the filtration cycle due to membrane fouling. Membrane fouling is often pronounced as the feed concentration increases. In addition, polymeric membranes are subject to compression effects as the ΔP_T is increased, which further adversely affects the permeate flux. Thus, for many polymeric membranes, large ΔP_T values cannot be applied, necessitating the use of extended recirculation or diafiltration. As a practical consequence of this, ultrafiltration systems using polymeric hollow-fiber membranes can be utilized only in the clarification of pressed, prefiltered juices. Even in these applications, suspended solids can cause blockage of the flow channels, and severe membrane fouling can occur due to concentration effects as a result of continuous recirculation.

Open tubular metallic and mineral membranes offer the advantage of handling solutions with high viscosity and high in suspended solids, due to large circular flow channels. The disadvantage of ceramic or metallic membranes, however, is the relatively low packing density and higher initial membrane element cost. The $Zr(OH)_4$ or $Zr(OH)_4$–polyacrylic acid mem-

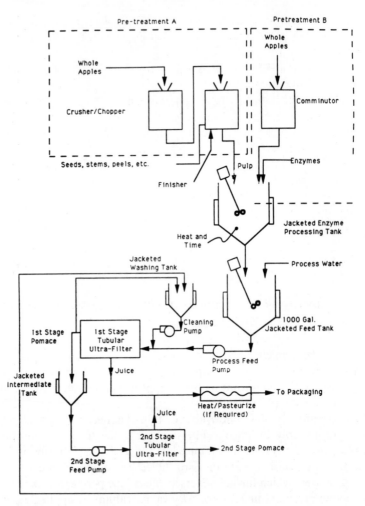

Figure 8.22. Ultrapres™ operation process sequence for apple juice clarification using Carre metallic membranes (Carre 1986).

brane layers on porous stainless steel were formed-in-place on the inner surface of the sintered stainless steel tubes. Figure 8.22 shows the schematic of a typical industrial-scale ultrafiltration system for the filtration of apple puree to produce clarified apple juice.

Fouling was much less pronounced for the larger diameter tube (0.03 m) operating at high feed velocity and low juice recovery, and yielded higher flux than the smaller diameter tube (0.016 m). The membranes were cleaned-in-place with an acid/base wash consisting of 0.015% sodium carbonate (of pH 9) followed by a 0.02 M ammonium acetate buffer (of pH 4.8) and with an intermittent hot water flush. Excellent juice quality was obtained by UF at 50°C. Permeate flux increased with temperature up to 50°C and leveled off above 50°C. The Ultrapres™ system has also been reportedly tested with promising results for carrot and beet juice clarification/concentration (Swientek 1987).

Conventional apple juice processing requires about 8 h whereas with the Ultrapres™ system the processes of apple grinding through juice clarification can be completed in only 2 h. It is also claimed that the process lowers the apple processing costs by 10–15 cents per gallon of juice. Moreover, the UF system also produces a clear juice full of flavor and exhibits a transmission haze of 0.17%.

8.3.3. Processing of Cranberry Juice with Ceramic Membranes

Traditionally, cranberry juice processing involves the use of a variety of equipment listed below (either alone or in combination):

1. Plate and frame diatomaceous earth (DE) filters
2. Rotary vacuum with filter aid
3. Centrifuge separators
4. Decanters

The various steps in the traditional cranberry juice processing are (1) Crushing (2) Depectinization (3) Pressing (4) Filtration (5) Evaporation and (6) Bottling.

In the processing of cranberry juice, significant differences are observed with respect to fresh and frozen cranberries. These are:

1. Pectin causes the most fouling.
2. Frozen berries tend to stay frozen in the center during the depectinization process. This increases the amount of pectin entering the clarification unit.
3. Considerably lower fluxes result when frozen berries are used.

The effect of pectin concentration on the performance of the membrane can be described in terms of the transport resistance (Venkataraman, Silverberg and Giles 1988):

1. The contribution in terms of resistance to transport due to pectin at 100 ppm concentration in the feed (or retentate) was 15.6%.
2. The contribution in terms of resistance to transport due to pectin at 1000 ppm concentration in the feed (or retentate) was 22.2%.

3. The contribution in terms of resistance to transport due to pectin at 10,000 ppm concentration in the feed (or retentate) was 46.6%.

The resistance model given below was used to determine the total resistance.

$$\frac{1}{J} = \left(\frac{R_m + R_s + R_p}{\Delta P_T}\right) \qquad (8.2)$$

where,

J = flux
R_m = resistance due to membrane
R_s = resistance due to suspended solids
R_p = resistance due to pectin
ΔP_T = transmembrane pressure

A practical case study involving the use of ceramic microfilters in the clarification of cranberry juice was also described by Venkataraman, Silver-

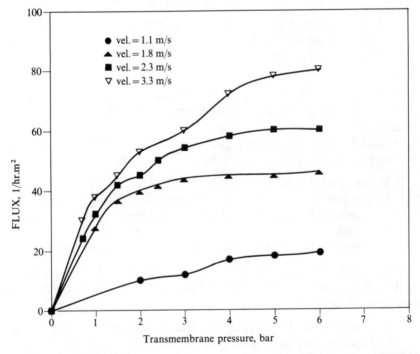

Figure 8.23. Flux versus transmembrane pressure behavior with Ceraflo 0.2 micron membrane in the clarification of cranberry juice (Venkataraman, Silverberg and Giles 1988).

berg and Giles (1988). For this application, a 0.45 μm alumina microfilter was selected. Experiments were also performed with a 0.2 μm and a 1.0 μm microfilter. Figure 8.23 shows the dependence of flux on ΔP_T at various recirculation rates for the 0.2 μm microfilter. Initial testing was performed using single-tube (inside diameter = 3 mm) modules, whereas industrial-scale tests were performed using multilumen element modules (channel diameter approximately 2 mm).

It is evident from Figure 8.23 that the flux is highest at a transmembrane pressure of 3 bar, at the highest recirculation rate. Similar measurements were performed on the 1.0 μm ceramic microfilter. The results are shown in Figure 8.24. These are essentially identical to the data obtained with the 0.2 μm membrane. However, the permeate quality was inferior to that observed with the 0.2 μm microfilter.

The effect of cross-flow velocity on flux is shown in Figures 8.25 and 8.26 for 0.2 μm and 1.0 μm microfilters, respectively. These data show that the slope values are almost the same for the 0.2 μm and the 1.0 μm membrane. From these and other measurements, the optimum pore diameter was

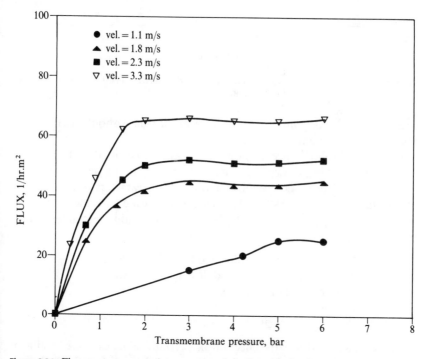

Figure 8.24. Flux versus transmembrane pressure behavior with Ceraflo 1 micron membrane in the clarification of cranberry juice (Venkataraman, Silverberg and Giles 1988).

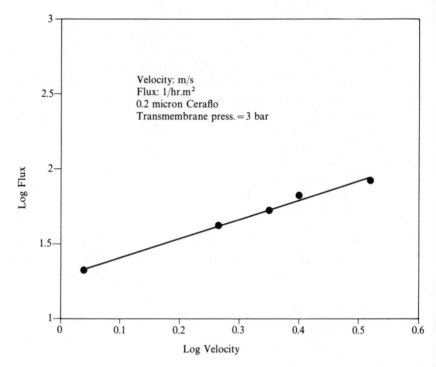

Figure 8.25. Log flux versus log velocity for depectinized cranberry juice (0.2 micron membrane) (Venkataraman, Silverberg and Giles 1988).

determined to be 0.45 μm, which maximizes the flux without sacrificing the desired permeate quality.

The effect of solids concentration on flux was also evaluated. The data for 0.2 μm Ceraflo™ are shown in Figure 8.27. Similar results were obtained for the 0.45 μm with cross-flow velocity and ΔP_T as parameters (Figure 8.28). It is clear that there is no substantial gain in flux above a ΔP_T value of about 3.5 bar. The optimal operating conditions for the clarification of cranberry juice are summarized below:

Inlet pressure to the filter module = 5.75 bar
Outlet pressure at the module exit = 1.25 bar
Avg. transmembrane pressure   ≃ 3.5 bar
Temperature   = 45°C
Cross-flow velocity   = 7 m/s.

Figure 8.28 also shows the effect of the percentage of suspended solids on flux under the above-listed optimal conditions for 0.45 μm Ceraflo™ alumina membrane.

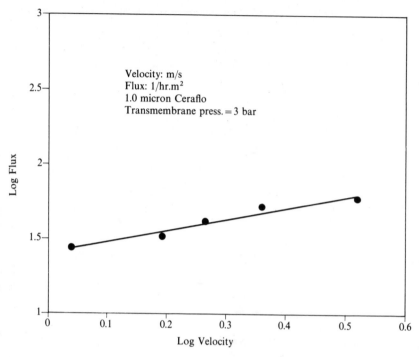

Figure 8.26. Log flux versus log velocity for depectinized cranberry juice (1 micron membrane) (Venkataraman, Silverberg and Giles 1988).

The procedure used for membrane regeneration involved the use of hot water to remove the solids previously diafiltered to low Brix. Caustic soda and sodium hypochlorite were added to the feed tank until an active chlorine level between 500 and 1000 ppm was obtained at a pH of 11. The solution was recirculated at 55°C for approximately 1 hour. A final rinse with deionized water was performed. The total downtime for cleaning was less than 90 minutes.

Amongst the numerous advantages in using ceramic microfilters, the primary ones are:

Quality of End Product From a quality standpoint, using a 0.45 μm ceramic (e.g. alumina) microfilter, the quality of the clarified cranberry juice (without color or flavor loss) was less than 1 NTU as compared to more than 10 NTU typically obtained with DE filters. The use of ceramic membranes also facilitates reduced holdup volume and can significantly reduce downtime.

Health and Environmental Factors For a fruit juice clarification plant using ceramic membrane filters, there is no need to be concerned about

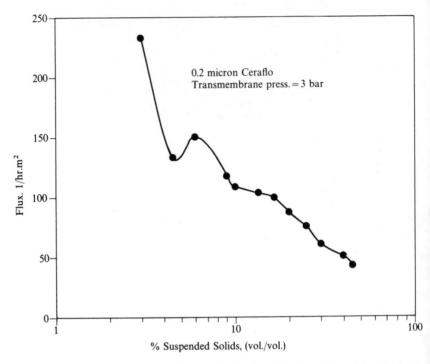

Figure 8.27. Effect of solids concentration on flux for depectinized cranberry juice (Venkataraman, Silverberg and Giles 1988).

silicosis. The problems related to current or future DE waste disposal are also eliminated. Thus, the ceramic membrane process offers an alternative technology that poses no health or environmental concerns.

Cost Comparison Venkataraman, Silverberg and Giles (1988) have also described the results of the cost comparison for the processing of a single-strength cranberry juice using a 0.45 μm alumina microfiltration system versus a DE filter system. For the base year 1987, the following assumptions were made:

1. Process time   100 days/year
2. Productivity   190 m^3 single-strength juice
3. Operation   16 hours/day
4. Labor and overhead cost $30/hour
5. DE cost $0.3/kg
6. Disposal cost $5,000 rental/handling + $100/load
7. Maintenance   3% of capital expense/year
8. Processing cost (less fruit) $100/$m^3$ of juice

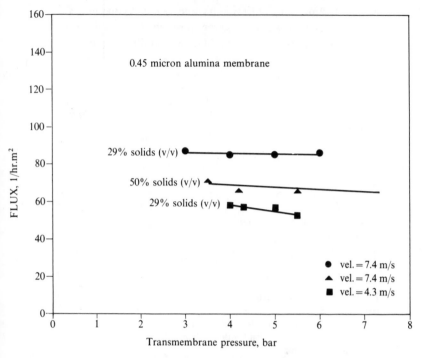

Figure 8.28. Flux versus transmembrane pressure behavior with Ceraflo membrane in the clarification of cranberry juice (Venkataraman, Silverberg and Giles 1988).

The annual operating costs along with the operating savings obtained with the ceramic membrane filtration system are given in Table 8.3.

8.3.4. Clarification of Strawberry and Kiwifruit Purees

The clarification of strawberry pulp obtained by grinding and sieving (mesh size 0.55 mm) was achieved using a multichannel ceramic membrane (Membralox® zirconia membrane) with a pore diameter of 0.1 μm (Tarodo 1990b). At 30°C, with an average ΔP_T of 2 bar and a tangential velocity of 5.6 m/s, a flux of about 50 L/h-m² was obtained at a concentration factor of 2.

Pectins were very strongly retained with a retention coefficient of about 85% whereas the retention of sugars was only 6.5%. The quality of ultrafiltrate was excellent, displaying a limpid and shiny appearance. The retention of coloring matter (largely attributed to anthocyanins) was less than 5%.

The filtration of kiwifruit puree was also performed with 0.1 μm Membralox® zirconia membranes. A flux value of about 65 L/h-m² was obtained

Table 8.3. Processing of Cranberry Juice: Annual Operating Costs and Total Operating Savings Using Ceramic Membrane Filtration Systems (Venkataraman, Silverberg and Giles 1988)

Cost Items*	Cost Per Year[†] ($)	
	DE Filter Press	Membrane Filter
DE costs	45,818	0
DE disposal costs (incremental)	5,750	0
Juice loss	115,040[‡]	58,500
Juice rework	20,000[§]	0
Maintenance	4,000	15,000
Energy	1,527	15,360
Cleaning chemicals (incremental)	0	1,750
Productivity losses	120,384[‖]	0
Labor costs	48,000[#]	0
Total operating costs	$360,519	$90,610
Total annual operating savings		$269,909

* All cost value data derived from actual operations
[†] The typical pay-back period for this system is 2–3 years
[‡] 143.8 m^3 at $800/m^3$
[§] 22.7 m^3/week at $52.8/m^3$
[‖] 1 hour/day, 11.4 m^3 at $105.6/m^3$
[#] 2 man-years

under the operating conditions described above for the clarification of strawberry puree. The retention of pectins was as high as 98%, with a low sugar retention of only about 5% (Tarodo 1990b). The permeate appearance was shiny and limpid displaying a light green color (Lozano et al. 1986, Tarodo 1990b).

A 50:50 mixture of strawberry and kiwifruit purees was prepared and filtered with the same zirconia membrane described above, under identical operating conditions used in the processing of the individual fruit pulps. A flux value of 60 L/h-m^2 was realized with 97% retention of pectins and 7% sugar retention. The filtrate was, once again, limpid, shiny and strongly colored due to the weak retention of anthocyanins ($\sim$ 16%). The results are summarized in Table 8.4. These applications, although yet to be commercialized, are under active development.

Table 8.4. Retention Rates During Microfiltration of Strawberry and Kiwifruit Purees (Tarodo 1990b)

Puree	Total Dry Matter (%)	Proteins (%)	Sugars (%)	Pectins (%)	Anthocyanins (%)
Strawberry puree	10.6	67	6.5	85	2.8
Kiwifruit puree	23.2	49.7	2.9	97.7	—
Mixture 50:50	24.4	64.1	7	96.7	16.1

$\Delta P_T = 2$ bar, tangential velocity $= 5.6$ m/s, temperature $= 30°C$
All % values in (wt./wt.)

8.4. APPLICATIONS OF INORGANIC MEMBRANES TO CONCENTRATE PROTEINS IN FOOD INDUSTRY

For applications involving protein concentration, membrane processes, particularly, ultrafiltration provide an attractive alternative to conventional technologies such as evaporation. Amongst the various membranes available on the market, inorganic UF membranes can uniquely handle the high solids concentration encountered in the production of concentrated protein products such as egg protein and soy-milk protein.

Currently, there are a few industrial plants in operation for the concentration of egg white, whole egg proteins and soy-milk (Merin and Daufin 1990, Berot 1989). Although ceramic membranes have yet to find wide use in this area, they appear to show considerable promise for a larger market share in the next few years.

At the time of this publication, there appears to be no published reports on the use of inorganic membranes for concentrating egg proteins. The available information is very scanty. Ceramic membranes with a molecular weight (MW) cutoff of 20,000 (e.g. Tech Sep M4, ZrO_2 on carbon membrane) are reportedly used to concentrate solids up to 32–35% from their initial 11–12% solids content (Merin and Daufin 1990). For the concentration of whole egg proteins, a 70,000 MW cutoff membrane is used (e.g. Tech Sep M1 membrane). In this application, a final liquid product with about 50% solids was obtained from a feed containing 24% solids.

8.4.1. Concentration of Soy-milk Proteins

Ultrafiltration membranes have been commercially used for the concentration of soy-milk proteins (Berot 1989, Berry and Nguyen 1988). The use of UF membranes allows the concentration of soy-milk proteins to high solids content compared to conventional concentration techniques such as evaporation.

Ultrafiltration, with or without diafiltration, allows the efficient removal of oligosaccharides (e.g. raffinose and stachyose) which are highly undesirable. Oligosaccharides cannot be easily broken down in the human digestive system and thus could cause flatulence in the lower intestines.

In a recent report, the use of ceramic UF membranes on the industrial scale for the concentration of soy-milk solids from 4% to 23% (wt./wt.) was described (Berot 1989). The industrial plant was equipped with $57 \, m^2$ membrane area and utilized Tech Sep M4 membranes (MW cutoff = 20,000). The operating temperature was in the range 40–50°C.

In the initial concentration step, the solids were concentrated from about 4% to 12.5%. The flux values averaged about 50–60 L/h-m^2 for this step. This was followed by diafiltration to remove small molecules while further concentrating soy-milk solids. The average flux values in this step were in the range 85–90 L/h-m^2. In the final concentration step, the soy-milk solids content was raised to 23%. Flux values of about 30 L/h-m^2 were realized. The apparent viscosity of the final concentrate was approximately 100 mPa-s.

8.5. CLARIFICATION OF FERMENTED ALCOHOLIC BEVERAGES USING INORGANIC MEMBRANES

The manufacture of wines is an extremely complex process typically requiring more than ten processing stages (Kilham 1987, Rios et al. 1989). In this complex process there are several stages where clarification is needed before the material can flow into the next processing stage. The five most common clarification/filtration stages are:

1. Clarification and biological stabilization of musts
2. Centrifugation after first racking of young wine
3. Diatomaceous earth filtration after first racking of young wine
4. Prefiltration with depth filters
5. Clarification and biological stabilization of unprocessed wine by conventional dead-end membrane filtration *or*
6. Clarification and stabilization of unprocessed wine using the relatively recent application of cross-flow membrane filtration

In the production of white wines using cross-flow membrane filtration, steps (2)–(4) can be replaced, permitting the direct filtration of wine after the clarification and stabilization step. On the other hand, red wines can be produced using cross-flow membrane filters directly after step (1).

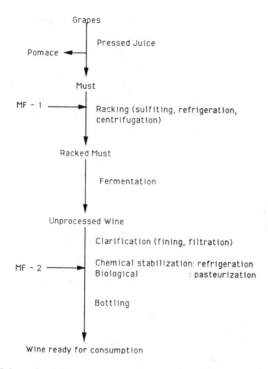

Figure 8.29. Schematic of the steps involved in the production of wine (Rios et al. 1989).

Figure 8.29 illustrates the modern wine-making process and shows the various stages where membrane filtration is used. The number of stages using membrane filtration will generally depend on the characteristics of grapes used as well as on the desired properties of the ready-to-consume final product. In addition to filtration steps, the pretreatment, stabilization and clarification steps are also essential with some of these occurring more than once during the entire wine-making process (Rios et al. 1989).

As is evident from Fig. 8.29, microporous filtration follows depth filtration and constitutes the first filtration phase, whereas the final filtration phase (polishing) is accomplished using cartridge filters often to produce dry stable wines. Membrane filtration is used after the polishing step, where sterile filtration is required for wines containing fermentable sugars or which undergo a "malolactic" fermentation.

Inorganic membranes have been used in the clarification of fermented alcoholic beverages such as young wine, beer and vinegar for the past 5–10 years (Poirier et al. 1984, Castelas and Serrano 1990, Merin and Daufin 1990). Relatively high flux values of about 150–250 L/h-m^2 are obtained with white wine and somewhat low flux values of only 50–100 L/h-m^2 are obtained for th clarification of red wine in small-scale to medium-scale industrial plants (Mietton–Peuchot 1985, Guibaud 1990). Figure 8.30 shows the layout of a medium-scale wine manufacturing plant using ceramic membranes. It is also widely recognized that for inorganic membranes to penetrate into large-scale wine producing plants, higher flux values are needed along with a better understanding of the effect of wine composition (which varies throughout the year) on the quality of wine produced by filtration.

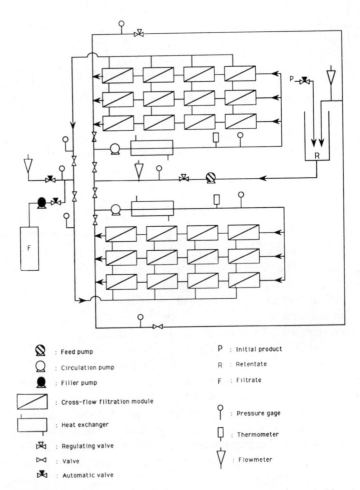

⊗	: Feed pump	P	: Initial product
Ọ	: Circulation pump	R	: Retentate
⬤	: Filler pump	F	: Filtrate
▱	: Cross-flow filtration module		
▭	: Heat exchanger	ⵁ	: Pressure gage
⋈	: Regulating valve	⊔	: Thermometer
⋈	: Valve		
⋈	: Automatic valve	▽	: Flowmeter

Figure 8.30. Schematic of an industrial-scale cross-flow microfiltration plant for the production of wine (Rios et al. 1989).

8.5.1. Colloidal and Noncolloidal Wine Components

Considerable variations in the filtration characteristics of wines are typically the result of complex mixtures of suspended and colloidal matter present in natural grape juice. Additional variations occur during the alcoholic and malolactic fermentation as well as during fining treatments (Harris 1964, Millipore Corporation 1974, Boulton 1984).

There are various colloidal and noncolloidal particles typically found in most wines. These are listed in Table 8.5, along with their approximate size range. They originate from the grapes, fermenting yeasts, bacteria and moulds. A knowledge of the wine composition is essential since it often determines the degree of clarification needed. Indeed, for some varieties of well-aged dry red wines, no filtration may be necessary (Kilham 1987), whereas other wines such as unprocessed red wines (e.g. wines made by fermentation under regulated pressure using bubble caps, chaine classic, etc.) need substantial clarification (Rios et al. 1989, Mignonac et al. 1985).

Microporous membrane filters are generally very effective in removing colloidal particles, particularly those which deform under high shear as is the case with cross-flow filtration. However, they also suffer from a disadvantage due to a low dirt-holding capacity and may plug relatively rapidly in the presence of high colloids concentration as compared to depth filters. In view of this situation, depth filters such as cartridge filters are often placed ahead of membrane filters to deal with the high colloidal content of many wines.

Table 8.5. Colloidal and Noncolloidal Particles Typically Found in Wines (Kilham 1987)

Particle Size (μm)	Particle Description
0.01–0.1	Peptides
	Proteins
	Polysaccharides
	Gums
	Dextrins
	Pectins
	Condensed tannins
	Leucocyanins and anthocyanins
	Phenolic compounds
0.5–1.0	Tartrate crystals
1–3	Yeasts (*Saccharomyces sp., Hensenula, Torulopsis, Debaromyces, Candida, Kluyveromyces, Pichia* and *Brettanomyces*)
0.5–7	Bacteria (*Leuconostoc, Acetobacter*, etc.)
50–100	Diatomaceous earth, fibers and debris

8.5.2. The Role of Filtration in the Process of Wine Making

Membrane filtration has played a significant role in the 10 billion gallon per year wine industry. The last half a century has witnessed significant technological advances in the process of wine making. Membrane filtration was introduced about 25–30 years ago in the process of wine making. The use of cross-flow filtration in the bulk processing of wines is under development, primarily in France, Germany, Italy and Australia and to a lesser degree in the U.S.

The clarification and biological stabilization of unprocessed wines, such as red and white wines, with membrane filtration is defined as the separation process that produces a clear and brilliant filtrate which is chemically and biologically stable. The retentate is a semisolid phase consisting of rejected particles, yeast, bacteria and a suspension of colloidal material. The membrane filtration process proceeds until a nearly solid residue is obtained (from the retentate) and is typically only about 1–2% of the feed. This suggests that the process recovers almost 98–99% of the feed as filtrate.

The principal purpose of using microporous membrane filters is their ability to remove yeast and bacterial cells. The removal of microorganisms by a membrane filter can also reduce the amount of SO_2 or ascorbic acid in wine (Mignonac et al. 1985, Kilham 1987). Cross-flow filtration is particularly desirable as it offers the possibility of producing sterile wines without extensive pretreatment and prefiltration. Utilizing such an operational mode, traditional processing steps such as racking, centrifugation, diatomaceous earth filtration and cartridge filtration can be partially or totally eliminated.

8.5.3. A Comparative Evaluation of the Various Filtration Processes Used in Wine Clarification

There are several different types of filtration needs in the complex process for the production of wines. The processing of wine, which involves the clarification and stabilization steps, traditionally begins with fining followed by successive filtration operations using diatomaceous earth filters and pleated cartridge depth filters (or filter pads). The advantage of using cross-flow membrane filters, as indicated earlier, is their ability to eliminate these steps and produce clarified, sterile wine ready for bottling. Inorganic cross-flow membrane filters have taken this technology even a step further. They allow the repeated application of chemical cleaning using strong acids and alkaline solutions to fully regenerate the membrane and offer long operating lifetimes (typically 5 years or more). The flux values are also generally higher than conventional polymeric membranes with reduced fouling, thus increasing the time interval between successive cleaning cycles (Poirier et al. 1984, Galaj et al. 1984).

Inorganic membranes also typically require substantially lower quantities of pretreatment chemicals or other consumables as compared to polymeric membrane filters, diatomaceous earth filters or pad filters (Rios et al. 1989).

8.5.4. Influence of Process Variables on Cross-flow Filtration: Clarification of White and Red Wines

The prediction of filtration performance through quantitative mathematical description of the effect of operating parameters on flux and fouling has been a major difficulty for many years. The literature on the modeling of wine filtration is scanty. In spite of the recent work of Belleville et al. (1990) which suggests a possible polysaccharide role in fouling, the phenomena of fouling and flux decline are not clearly understood, even for conventional filters such as diatomaceous earth, composite pad filters or synthetic polymeric membrane filters (de la Garza and Boulton 1984).

Membrane Pore Diameter The choice of membrane pore diameter for microporous cross-flow filtration depends on the ability of the membrane structure to yield reasonably high filtrate flux together with a minimum retention of organoleptic and coloring substances. Various pore diameters in the range 0.1–1.5 μm have been considered in laboratory and industrial trials. However, the wide acceptance of 0.2 μm membranes is due to the requirement that the retention of particles and colloids as well as microorganisms by the microporous membrane must be excellent. Additionally, pore plugging and/or membrane fouling are less severe with the smaller pore membranes which are also capable of handling high solids concentration. The presence of colloids can cause turbidity problems whereas the elimination of microorganisms such as bacteria and yeasts is necessary to ensure sterility.

The use of 0.2 μm membranes also ensures that the organoleptic characteristics due to polysaccharides, polyphenols and alcohols of the clarified wine are not significantly affected. Typical data demonstrating the effect of pore diameter on flux are shown in Table 8.6. These data obtained for the clarification of red wine (Poirier et al. 1984) clearly show that 0.2 μm membranes have the best overall performance over the range of pore diameters studied.

System Operating Parameters There are many system parameters that may need to be optimized in order to obtain clarified and sterile wine of consistently acceptable quality. The effects of cross-flow velocity, transmembrane pressure (ΔP_T) and temperature on the clarification of red wine are illustrated in Table 8.7 whereas Figure 8.31 shows the effect of concentration on flux.

Table 8.7 shows that flux increases linearly with cross-flow velocities in the range 2.5–4.7 m/s. Chemical analysis indicated that the quality of filtrate was

Table 8.6. Clarification of Red Wine: Effect of Membrane Pore Diameter on Flux (Poirier et al. 1984)

Pore Diameter* (μm)	Bacteria Count[†] in Permeate	Flux (L/h-m²)
0.2	0	83
0.45	0	87
1.2	10–100	73

Cross-flow velocity = 4.7 m/s, $\Delta P_T = 3$ bar, temperature = 20°C
* Membralox® Al_2O_3 membranes
[†] Number of microorganisms/100 ml

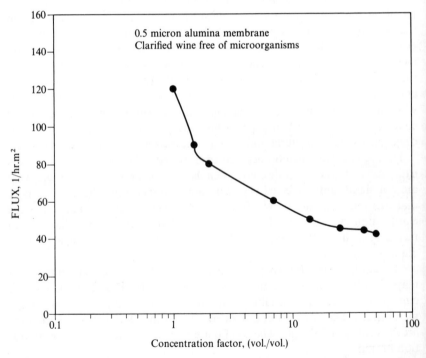

Figure 8.31. Effect of concentration factor on flux in the clarification of wine (Poirier et al. 1984).

Table 8.7. Clarification of Red Wine: Effect of Cross-flow Velocity, Transmembrane Pressure and Temperature on Flux (Poirier et al. 1984)

Cross-flow Velocity (m/s)	Transmembrane Pressure (bar)	Temperature (°C)	Bacteria* Count in Permeate	Flux (L/h-m²)
2.5	3	20	0	55
3.1	3	20	0	65
3.9	3	20	0	75
4.7	3	20	0	95
4.7	1	20	0	65
4.7	2	20	0	75
4.7	3	20	0	85
4.7	5	20	0	85
4.7	7	20	0	85
4.7	9	20	0	85
4.7	3	10	0	60
4.7	3	20	0	95
4.7	3	30	0	120
4.7	3	40	0	150

Pore diameter = 0.2 μm Membralox® Al_2O_3 membranes
* Number of microorganisms/100 ml

not adversely affected and that no microorganisms (e.g. bacteria) were detected in the permeate. A cross-flow velocity value of about 5 m/s is typical in order to limit the pressure drop to within 1 bar. The effect of transmembrane pressure on flux and properties of clarified wine are significantly more complex. At low transmembrane pressures (e.g. up to 3 bar), flux increases linearly suggesting that the operation belongs to the pressure-controlled regime. At higher ΔP_T values, the flux usually levels off (e.g. 75 L/h-m²). However, in this particular case, high transmembrane pressures are undesirable since the organoleptic characteristics are also modified apart from the high energy consumption.

Although the clarification of red wine at higher temperatures (up to 40°C) resulted in a linear increase in flux (Table 8.7), temperatures higher than about 20°C are not acceptable due to the negative effect on the organoleptic properties (Poirier et al. 1984).

As illustrated in Figure 8.31, an increase in concentration over time for the semi-batch operation typically results in flux decline. This may be expected since an increase in the concentration of colloids on the membrane surface, under a given set of conditions, can lead to concentration polarization and fouling. Although flux decline is typical, the quality of wine produced is of acceptable quality with sterility. In practice, with the continuous intake of fresh wine, the incremental increase in feed concentration over a given period of time is quite small. This helps to maintain the filtration performance at a relatively steady value with a nominal rate of flux decline. Further, the use of backflushing helps smoothen the operation by avoiding excessive pressure buildup by periodically dislodging accumulated particles on the membrane surface.

8.5.5. Vinegar Filtration with Ceramic Membranes

The fermentation of diluted alcohol to produce vinegar results in a turbid liquid due to the presence of acetobacters. The vinegar concentration is typically 10–15% by weight. Continuous cross-flow microfiltration is used to obtain a clarified product by removing acetobacters in the concentrate along

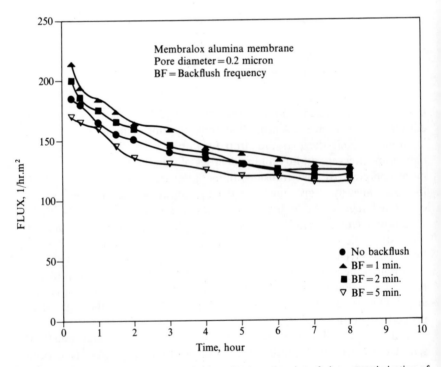

Figure 8.32. Performance of ceramic membranes in the processing of vinegar: optimization of backflush (Francoisse and Hermia 1989).

with bacteria. The filtration is continued for 2–4 weeks depending on the desired concentration factor. For instance, with microporous polypropylene membranes, up to 30 days were required to achieve a concentration factor of 60 (Ripperger and Schulz 1986).

The use of inorganic membranes such as alumina and zirconia in the purification of vinegar was also reported (Merin and Daufin 1990, Francoisse and Hermia 1989). The literature on this application is very scanty. From the available information, it appears that 0.1 μm ZrO_2 and 0.2 μm pore diameter Al_2O_3 membranes are very effective in turbidity reduction and produce clear filtrates with turbidity values in the range 0.2–0.8 NTU.

Using 0.2 μm Membralox® alumina membranes, high flux values in the range 115–130 L/h-m^2 were obtained. These data are shown in Figure 8.32 (Francoisse and Hermia 1989). Backflushing was employed to maintain high flux values and minimize fouling. The optimum ΔP_T and cross-flow velocity were found to be around 2 atm and 5 m/s, respectively.

Similar studies with 0.1 μm pore diameter ZrO_2 membranes supported on α-Al_2O_3 gave somewhat higher flux values in the range 150–170 L/h-m^2. The

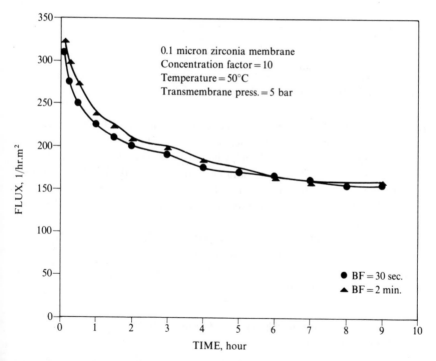

Figure 8.33. Performance of Membralox® ceramic membranes in the purification of vinegar (Francoisse, Hermia 1989).

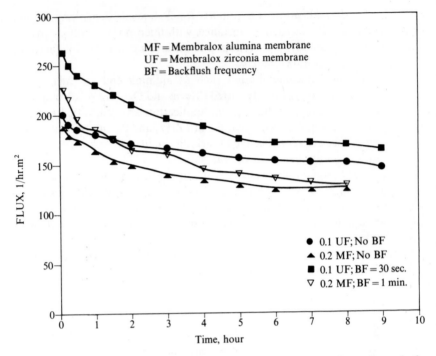

Figure 8.34. Performance comparison of 0.1 micron versus 0.2 micron ceramic membranes in the processing of vinegar (Francoisse and Hermia 1989).

performance of a 0.1 μm membrane is shown in Figure 8.33 (Francoisse and Hermia 1989). For this membrane, periodic backflushing at 30 s intervals gave the highest flux of 170 L/h-m^2, whereas for intervals of 1–5 min the flux value of about 150 L/h-m^2 was obtained. In all cases, the quality of filtrate was excellent, with turbidity values below 1 NTU. A performance comparison for the 0.1 and 0.2 μm membranes is shown in Figure 8.34 (Francoisse and Hermia 1989).

Zirconia membranes on carbon support are also reported for the commercial production of vinegar (Merin and Daufin 1990). The molecular weight cutoff of the composite ZrO_2 membrane was 70,000 (Carbosep® M1 membrane). Flux data are not reported.

8.5.6. Bacteria Removal with Inorganic Microfilters to Produce "Cold" Sterile Beer

In beer pasteurization, although the traditional filtration process is effective in virtually eliminating all yeast cells and microorganisms (e.g. lactobacillus) from passing into the filtrate (product phase), it does not guarantee 100% bacteria retention. Another problem with the traditional pasteurization process is the high thermal load which can adversely affect organoleptic qualities attributed to organic flavor-producing compounds. Bacteria removal by cross-flow microfiltration is being evaluated as an alternative to the traditional beer pasteurization process involving heat treatment, to ensure sterility prior to bottling. It is estimated that about 85% of all beer produced is clarified using filtration (Reed 1986). The sequence of operations for the clarification of beer is shown in Figure 8.35.

In addition to bacteria removal, inorganic cross-flow microfilters can also improve the quality of the clarified beer. This may not necessarily be advantageous since it can significantly alter the color and taste due to some loss of bitter-flavored compounds. Bacteria removal with 0.2 and 0.5 μm pore diameter Membralox® alumina membranes was recently reported by

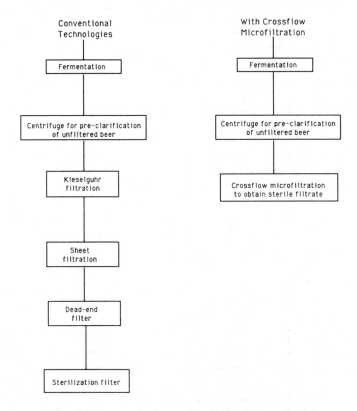

Figure 8.35. The sequence of operations for the clarification of beer.

Tragardh and Wahlgren (1990). The filtration performance was evaluated in pilot-plant tests by operating under constant transmembrane pressure (CTP) conditions. This is similar to the Bactocatch process, illustrated in Section 8.2.1, for bacteria removal from milk. Various CTP conditions up to 1 bar were used. However, no flux data are reported. Although the measurements were made at 20°C, filtration performance at low temperatures (around 0°C) is expected to be better, particularly from the standpoint of retention of the organoleptics responsible for color and taste. The optimum operating temperature is brewery dependent.

The product characteristics obtained with 0.2 μm membranes were unacceptable due to high color retention (about 30%) and protein retention (approximately 12%). On the other hand, with 0.5 μm pore diameter membranes the color retention was only about 3% with insignificant protein retention. No bacteria were detected in the filtrate.

These preliminary results appear encouraging and suggest that, with further performance improvements, the cross-flow microfiltration technique for the production of sterile beer may become economically and technically competitive with other filtration technologies. The efficiency of the membrane cleaning process also needs to be improved. Membrane fouling was quite severe, probably as a result of pore blockage with bacteria and/or proteins which could not be easily removed in the chemical cleaning step.

8.5.7. Recovery of Beer by Clarification of Tank Bottoms

There are four main steps in the manufacture of beer. These are:

1. Extraction of malted barley and/or other materials with water
2. Heating the extract to boiling
3. Cooling the extract and fermenting with yeast, and
4. Clarification

The fermented mass is next centrifuged to recover yeast. The beer obtained from the supernatant fluid can be filtered at 0–4°C using ceramic membranes and stored at nearly 0°C in cold conditioning tanks. Filtration at low temperatures also ensures product stability while minimizing the loss of essential components (including alcohol) responsible for appearance and taste. A schematic of a typical cross-flow microfiltration plant for the processing of beer is shown in Figure 8.36 (Meunier 1990).

"Tank bottoms" is a coagulated liquid extract typically consisting of suspended solids, collagen finings, solutes and yeast. The possibilities for cross-flow microfiltration in beer tank bottoms processing are based on the potential for very significant savings in kieselguhr, a variety of diatomaceous earth commonly used in the brewing industry. The use of ceramic membranes will also result in at least 10% reduction in the volume (and hence cost) of effluent discharge (Finnigan, Shackleton and Skudder 1987).

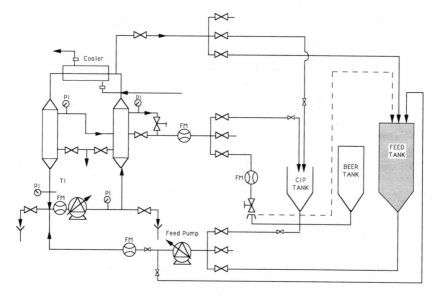

Figure 8.36. Schematic of a pilot-scale cross-flow microfiltration plant for the processing of beer, membrane area approximately 8 m² (Meunier 1990).

Low-temperature membrane filtration in the presence of collagen finings is limited by adsorptive fouling. Consequently, high concentrations of solids in the retentate cannot be achieved. In the absence of finings, however, higher concentrations (about 25% by wt.) are achieved thereby permitting higher beer recovery from tank bottom extract.

The use of 0.12–1.8 μm pore diameter Membralox® Al₂O₃ membranes for the recovery of beer from beer yeast and/or from cold conditioning tank bottoms (CCT) in the absence of finings has been reported (Finnigan, Shackleton and Skudder 1987, Meunier 1990, Walla and Donhauser 1990). Figure 8.37 shows typical flux values obtained during the microfiltration of beer yeast and tank bottoms to recover beer.

Figure 8.38 shows the effect of solids concentration on flux. The final concentration of 26% (by wt.), from an initial value ranging from 2–6% (by wt.), was obtained at an average ΔP_T of 4 bar. The steady-state flux after about 4 h of operation at the above final concentration was reported to be 20 L/h-m². Backflushing was found to be very helpful in minimizing the thickness of the fouling layer on the membrane. Significantly higher flux values of about 60 L/h-m² are expected for diafiltration of fermentation tank bottoms using ceramic membranes (Le 1987).

The selection of the optimum pore diameter is of great importance not only to permeate all the essential components such as alcohol (content about 5%

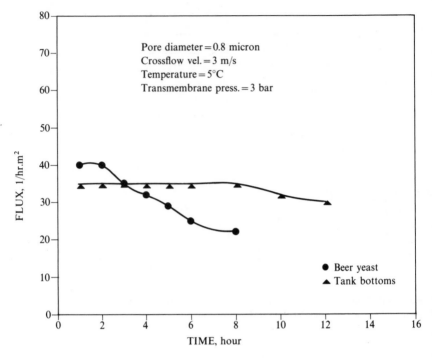

Figure 8.37. Performance of Membralox® alumina membranes in the processing of beer yeast and tank bottoms (Meunier 1990).

by volume) but also to ensure that the desired product clarity (haze value 8–12 Monitek units), color and specific gravity are obtained. The retentate of the filtration loop for beer recovery was found to contain high levels of proteins (about 50%) and may be used as an animal feedstock.

A conservative estimate of the potential savings achievable with ceramic filtration of beer as compared to conventional filtration systems using diatomaceous earth was reported (Shackleton 1987, Finnigan, Shackleton and Skudder, 1987). These are summarized in Table 8.8 and do not include additional savings in labor, handling and disposal costs resulting from the substantial to total replacement of kieselguhr with ceramic filters. Although the actual magnitude of savings is brewery dependent, the numbers clearly show areas where major savings can be accomplished.

8.6. INORGANIC MEMBRANES IN BIOTECHNOLOGY APPLICATIONS

Membrane filtration is used in many situations for the separation of cells from a fermentation broth and the recovery of valuable fermentation produc-

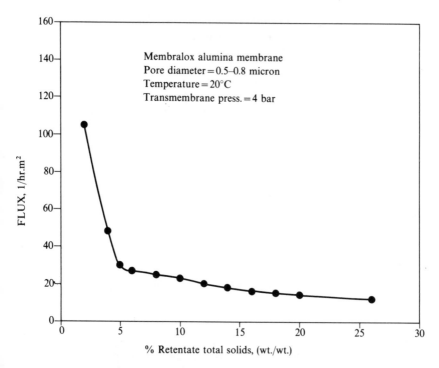

Figure 8.38. Microfiltration of beer tank bottoms flux versus retentate solids concentration (Finnigan, Shackleton and Skudder 1987).

ts such as proteins and antibiotics (Kroner et al. 1984, Hanish 1986, Datar 1984, Matsumoto, Kawahara and Ohya 1988). In other instances, microporous membranes are utilized in bioreactors where they serve as supports for enzymes and microorganisms and also allow the reaction and separation to occur in a single stage (Nishizawa et al. 1983, Hoffmann et al. 1985, Ripperger and Schulz 1986, Hoffmann, Scheper and Schugerl 1987). The available information in the open literature is limited due to the highly proprietory nature of processes involved in biotechnology applications. Therefore, the majority of information compiled here is derived from laboratory-scale or pilot-scale studies.

One drawback of laboratory-scale performance data with simulated fermentation broths is that these may not fully reflect the true behavior of fermentation broths on an industrial scale. Considerable membrane fouling may occur in industrial-scale filtration, depending on the characteristics of the membrane, operating configuration, transmembrane pressure and frequency and duration of backflush.

Table 8.8. Potential Savings for
Ceramic Filtration of Beer
(Finnigan, Shackleton and Skudder
1987)

Process Item	Savings ($ × 10^3)
Kieselguhr	100
Finings	100
Yield*	300
Effluent[†]	100
Total savings	600

Includes unfined and ex-cold conditioning
tank (i.e. beer solids concentrated in final
stages)
* Based on 1/2% of a million barrels
[†] Based on 10% of one million pounds
effluent charge per annum. This is brewery
dependent.

In biotechnology applications, product sterility and the ability of the filtration system to handle high-viscosity liquids are of paramount importance. The use of ceramic membranes is gaining considerable attention in many biotechnology applications due to their superior thermal and chemical resistance and high mechanical strength. These features permit easy sterilization, either by use of live steam or by using chemically active chlorine-based compounds such as hypochlorites. The separation of cells using microporous ceramic membranes has been reported in recent years by several groups (Taniguchi, Kotani and Kobayashi 1987, Bjorling and Olofsson 1988, Matsumoto, Kawahara and Ohya, 1988, Chun and Rogers 1988). Most ceramic membrane systems operate in the cross-flow filtration mode, whereas for the polymerics both cross-flow and dead-end filtration modes are used (Ripperger and Schulz 1986).

Many of the results achieved with ceramic cross-flow filters are indeed quite significant. For instance, Taniguchi, Kotani and Kobayashi (1987) reported the development of a new fermenter, where certain inhibitory metabolites (e.g. lactates) were continuously removed by a 0.2 μm alumina membrane. The process resulted in the complete retention of cells and allowed the realization of high cell concentration not achievable with conventional batch cultivation techniques.

Conventional cross-flow filtration is carried out with and without back-flushing where a small portion of the filtrate is forced back into the feed side as a means to limit flux decline. A novel cross-flow filtration technique developed by Alfa-Laval, where a constant transmembrane pressure (CTP) is maintained over the entire module, has been shown to improve overall flux and reduce fouling very substantially (see Chapter 5.6) in dairy and bio-technology applications (Bjorling and Olofsson 1988, Sandblom 1978, Gillot, Soria and Garcera 1990).

Another area of biotechnology-related applications is the production of pyrogen-free water or separation of low molecular weight organics from process streams. Certain types of ultrafiltration membranes are reportedly used to produce pyrogen-free water (Filson et al. 1991) or to perform the separation of low molecular weight organics such as amino acids and organic acids (Cueille and Ferreira 1990). Pyrogens (also known as endotoxins) are fever-inducing substances. Their molecular weight values range from 10,000 up to 200,000, or even higher. Pyrogens cannot be eliminated by autoclaving or microfiltration but have been successfully removed by synthetic polymeric UF membranes (Cheryan 1986). At this time, however, there appears to be no published description on the use of inorganic UF membranes in these applications. Small-pore capillary and tubular UF membranes produced by Alcoa/SCT, Corning Glass and Tech Sep may be successfully utilized in pyrogen removal applications.

8.6.1. Microorganism Separation and Cell Debris Filtration

Matsumoto, Kawahara and Ohya (1988) studied the cross-flow filtration of yeast with microporous ceramic membrane (1.4 μm Membralox®, Alcoa/ SCT, Tarbes) using several backflushing techniques. Concentration of yeast cells from a fermentation broth is an important operation in the production of beer. Periodic backflushing by forcing the filtrate backwards using com-pressed air proved most effective in preventing rapid flux decline. This is illustrated in Figure 8.39.

On the other hand, flux decline was rapid in the absence of backflushing. The dependence of flux on backflushing interval, time and filtration volume was studied. It was observed that the longer the backflushing interval (e.g. up to 5 min), the smaller the gain in flux was, resulting in a lower overall flux at the expense of higher filtrate volume. This is typical of high-flux micro-filtration membranes where the flux decline due to the accumulation of particle deposits on the membrane surface occurs within minutes. Periodic backflushing over shorter intervals (e.g. 1–3 min) is therefore desirable. A larger duration of backflush such as 5 or 10 s is generally not effective and only results in substantial loss of filtrate. The effectiveness of backflushing does not generally depend on duration but on the magnitude of force applied

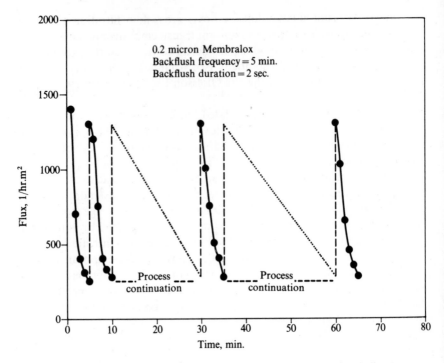

Figure 8.39. Filtration of yeast suspension: use of backflushing to minimize flux decline (Matsumoto, Kawahara and Ohya 1988).

on the fouling layer to dislodge the layer of particles and carry it into the retentate.

Another important aspect of application development for cell filtration and concentration using membranes is the relationship between cell concentration and filtration flux. It is evident that flux decreased with an increase in cell concentration. Backflushing was observed to be quite effective in minimizing flux decline with increasing biomass concentration. (Matsumoto, Katsuyama and Ohya 1987, Matsumoto, Kawahara and Ohya 1988).

Table 8.9 shows the comparative flux values for alumina membranes, and for polymeric membranes made of cellulose acetate (pore diameter 0.45 μm). Flux values with ceramic membranes were higher as compared to the polymeric membrane, under otherwise similar conditions. This is due to the lower degree of fouling experienced with ceramic membranes as compared to their polymeric counterparts. This may be explained on the basis of the common observation that backflushing in ceramic membranes is more effective as compared to that in polymeric membranes since higher back-

Table 8.9. Effect of Yeast Cell Concentration on Flux. (Matsumoto, Kawahara and Ohya 1988, Matsumoto, Katsuyama and Ohya 1987)

Membrane Type	Yeast Concentration (g/L)	Flux (m^3/m^2-h)	
		Yeast* Suspension	Fermentation[†] Broth
Al_2O_3, 1.6 μm	8.5	1–1.5	1.0
	30	0.4	—
Al_2O_3, 0.22 μm	8.5	0.5	—
	30	—	—
Cellulose acetate, 0.45 μm	8.5	0.3–0.4	0.2–0.3
	30	0.1	

* pH = 3–6, ΔP_T = 0.5–1 bar, temperature = 20°C
[†] pH = 6.0, ΔP_T = 0.5 bar, temperature = 20°C

pressure can be applied on the mechanically strong ceramic membranes which is responsible for dislodging particle layers and minimizing fouling. Fouled ceramic membranes could be regenerated by chemical cleaning using e.g. Ultrasil 54 (Henkel, Inc.) which contains active chlorine or by heating the membrane for 3 hours at 700°C (Matsumoto, Kawahara and Ohya 1988). Such a high-temperature cleaning procedure is, however, unlikely to be acceptable in industrial plants.

Chun and Rogers (1988) have described the use of a 0.2 μm ceramic microfiltration membrane in conjunction with a bioreactor for the production of ethanol/sorbitol. Traditional microorganisms such as yeast are conventionally used to produce ethanol. However, it was observed that under certain conditions, strains of *Zymomonas mobilis* can produce the low-calorie sweetener sorbitol along with ethanol (Viikari 1984a, b). Using equimolar mixtures of glucose and fructose (e.g. 100 g/L each), at a temperature of 35°C and a pH of 6.2, a high concentration of ethanol (75 g/L) and sorbitol (40 g/L) was obtained. The fouling of the ceramic membrane was minimal and contributed to a stable filtration operation permitting (a) high recovery of soluble products, namely ethanol and sorbitol.

Cross-flow filtration under conditions of constant transmembrane pressure (CTP), a novel technique recently developed by Alfa-Laval, was applied to the filtration of *E. coli* cell debris suspensions. Membralox® 0.2, 0.5 and 0.8 μm alumina membranes were tested for the recovery of proteins and lactase, and for the retention of cell mass (*E. coli* 30% by wt.). The test results are summarized in Table 8.10. It is evident that the 0.8 μm membrane gave the best performance with almost 100% recovery of protein and lactase, with

Table 8.10. Cell Debris Filtration Using Ceramic Membranes: Effect of Pore Diameter on Filtration Performance (Bjorling and Olofsson 1988)

Pore Diameter* (μm)	% Suspended Solids[†] (vol./vol.)		Total Protein (g/L)		Transmembrane Pressure[‡] bar	Flux (L/h-m²)
	Retentate	Permeate	Retentate	Permeate		
0.2	4	0	13.8	9.7	0.7–0.9	50
0.45	2.5	0.005	9.45	5	0.4–0.8	40
0.8	4	0.01	9.4	9.2	0.4–0.6	100

* Membralox® Al_2O_3 membranes
† E. coli cells
‡ Constant transmembrane pressure operation, cross-flow velocity = 7 m/s

quantitative retention of cells in the retentate. The 0.8 μm membrane retained cell debris as efficiently as the 0.2 μm membrane.

Flux values were significantly higher (viz. 110 L/h-m²) with the 0.8 μm membrane as compared to (about 30–40 L/h-m²) the 0.2 μm membrane. Further, the recovery of proteins and lactase was only about 80–85% with the 0.2 μm membrane as compared to greater than 95% with the 0.8 μm membrane. The performance of the 0.45 μm membrane was inferior relative to both the 0.2 and 0.8 μm membranes with respect to protein recovery and flux, and suffered from instability and severe pore blockage.

The superior performance under CTP conditions was also demonstrated for *E. coli* filtration from the observation of significant flux decline under conventional cross-flow filtration with backflush as opposed to a stable high flux value under the CTP mode. The retention of protein and lactase was also sharply higher.

8.6.2. Plasma Separation by Cross-flow Filtration Using Inorganic Membranes

Microfiltration membranes for plasma separation are required to pass the sterility test with live steam or in an autoclave. Inorganic membranes appear to be well suited due to their high thermal and chemical resistance and their ability to be fully regenerable on chemical or thermal treatment.

The treatment of many fatal diseases is made possible partly through the removal of toxic substances with novel techniques such as low-temperature plasma separation (Nose, Kambic and Matsuhara 1983, Ikeda et al. 1986). It is necessary to perform plasma separation and fractionation at low temperatures to prevent bacterial contamination and protein denaturation.

The properties of blood have a profound influence on the filtration performance in plasma separation (Malchesky et al. 1989). Properties such as blood and plasma viscosity, cell and protein concentration, cell type and size are of prime consideration. With plasma separation, the whole blood cell concentration (and hence the viscosity) increases substantially from inlet to outlet, which in turn leads to an increase in pressure drop. To achieve clinical standards, important parameters such as solute (protein) retention, transmembrane pressure ensure absence of hemolysis and plasma flux values are varied.

The application of membranes to separate plasma is also limited by fouling and gel polarization, phenomena common to almost all membrane-based liquid separations. Fouling occurs because serum proteins adhering to pore walls narrow the pore diameter (pore plugging) and/or accumulate on the membrane surface (Friedman et al. 1983, Randerson and Taylor 1983). Pore plugging is relatively minimal with inorganic membranes due to their narrow pore size distribution as compared to that in many synthetic polymeric membranes (Sakai et al. 1987, 1989, Ozawa et al. 1985). Further, fouled ceramic membranes such as alumina microfiltration membranes can be fully regenerated chemically (e.g. with a dilute sodium hypochlorite solution) or thermally.

Low-temperature plasma separation by cross-flow filtration with microporous glass membranes was studied using bovine serum and bovine blood feed solutions with a protein concentration of 65 kg/m^3 (Sakai et al. 1989). The bovine blood feed had a hematocrit content of 30%. In another study with microporous glass and alumina ceramic membranes, filtration measurements were performed on bovine plasma, bovine blood (protein concentration $= 65 \text{ kg/m}^3$, hematocrit $= 36\%$) and 0.1% aqueous bovine serum protein test solutions (Ozawa et al. 1985). The results for cross-flow filtration using bovine blood samples are summarized in Table 8.11.

For microporous tubular glass membranes, filtrate flux increased with ΔP_T up to 0.03 bar whereas for ceramic membranes flux increase was noted at pressures up to 0.1 bar. For the filtration of bovine plasma with ceramic membranes, filtrate flux was dependent on ΔP_T whereas for bovine blood it was independent of transmembrane pressure. This can be explained on the basis of concentration polarization being the limiting factor for bovine plasma filtration whereas for bovine blood, boundary-layer effects or mass transfer effects (i.e. gel polarization) control flux. Thus, by increasing cross-flow velocity or shear rate, flux may be increased (e.g. in the case of bovine plasma) as opposed to an increase in transmembrane pressure which is characteristic of pressure-controlled regime (e.g. in the case of bovine blood).

An important feature of cross-flow filtration with inorganic membranes is the absence of hemolysis (characterized by the degree of red cell damage) at

Table 8.11. Performance of Cross-flow Filtration in Low-Temperature Plasma Separation Using Ceramic Membranes (Sakai et al. 1989, Ozawa et al. 1985)

Membrane Type	Pore Diameter (μm)	Temperature (°C)	Transmembrane Pressure (bar)	Protein Retention (%)	Flux ($L/h\text{-}m^2$)
Symmetric	1	10	0.03	93	1
Microporous Glass	1	10	0.1	90	2
MPG MB10	1	37	0.03	~ 100	14
(Asahi Glass)	1	37	0.1	95	18
Composite	0.2	37	0.05	50	90
Alumina	0.2	37	0.1	50	110
(Alcoa/SCT					
Membralox®)	0.2	37	0.15	50	136

Glass membranes, wall shear rate = $2000\,s^{-1}$
Alumina membranes, wall shear rate = $1000\,s^{-1}$

higher pressures up to 0.133 bar (or higher) for glass membranes and ceramic membranes as compared to polymeric membranes such as polymethyl methacrylate, polyvinyl alcohol, polyethylene and cellulose acetate where the ΔP_T must be maintained below 0.07 bar to prevent hemolysis (Horiuchi et al. 1981, Malchesky et al. 1989, Smith et al. 1979).

Although plasma separation with composite microporous alumina membranes (0.2 μm Membralox®) does not suffer from hemolysis, the observed solute retention values are lower. For instance, the retention coefficients for albumin and total proteins were only 0.55 and 0.45, respectively. This may be attributed to the very thin separation layer on large-pore porous support. These composite membranes, however offer relatively very high flux values (e.g. > 100 L/h-m²) as compared to flux values in the range 10–20 L/h-m² obtained with microporous glass membranes.

On the other hand, microporous glass membranes (pore diameters ranging from 0.28 μm up to about 1.5 μm) gave a higher retention coefficient of 0.93 and 0.91 for albumin and total proteins, respectively. These values are comparable to those obtained with synthetic polymeric membranes (Malchesky et al. 1989, Ozawa et al. 1985, Sakai et al. 1989). The higher retention values can generally be attributed to the plugging of pores with proteins across the thickness (0.5–1 mm) of these symmetric membranes.

It may, however, be anticipated that lower pore diameter composite membranes in the range 0.01–0.1μm can realize protein retention values comparable to the symmetric glass membranes (Sakai et al. 1989) or synthetic polymeric membranes (Malchesky et al. 1989) with comparable or higher flux values.

REFERENCES

Alcoa/SCT. 1990. Membralox® Technical Brochure.

Attia, H., M. Bennasar and B. Tarodo de la Fuente. 1988. Ultrafiltration on a mineral membrane of biologically or chemically acidified milks (with varying pH) and of lactic coagulum (in French). Le Lait 68(1): 13–32.

Attia, H., M. Bennasar and B. Tarodo de la Fuente. 1991a. Study of the fouling of inorganic membranes by acidified milks using scanning electron microscopy and electrophoresis. I. Membrane with pore diameter 0.2 μm. J. Dairy Research 58: 39–50.

Attia, H., M. Bennasar and B. Tarodo de la Fuente. 1991b. Study of the fouling of inorganic membranes by acidified milks using scanning electron microscopy and electrophoresis. II. Membrane with pore diameter 0.8 μm. J. Dairy Research 58: 51–65.

Baker, R. W., E. L. Cussler, W. Eykamp, W. J. Korus, R. L. Riley and H. Strathmann. 1990. Membrane Separation Systems—A Research and Development Needs Assessment. U.S. Department of Energy, Office of Energy Research-Office of Program Analysis, Contract DE-AC01-88ER30133.

Belleville, M. P., J. M. Brillouet, B. Jarodo de la Fuente and M. Moutounet. 1990. Polysaccharide effects on cross-flow microfiltration of two red wines with a microporous alumina membrane. J. Food Sci. 55(6): 1598–1602.

Bennasar, M., D. Rouleau, R. Mayer and B. Tarodo de la Fuente. 1982. Ultrafiltration of milk on mineral membranes: improved performance. J. Soc. Dairy Technol. 35(2): 43–49.

Bennasar, M. and B. Tarodo de la Fuente. 1983. Moderate ultrafiltration of raw cold milk on mineral membranes: industrial experiments (in French). Lait 63: 246–65.

Bennasar, M., D. Garcera, J. Gillot and B. Tarodo de la Fuente. 1984a. A new mineral membrane (in French). Tech Lait 12: 25–31.

Bennasar, M., B. Tarodo de la Fuente, J. Gillot and D. Garcera. 1984b. Study of a new membrane for the ultrafiltration of milk and optimization of its geometry. In Filtra 84, 2–4 October, Paris.

Berot, S. 1989. Recent developments in the utilization of membrane techniques for the separation and purification of vegetable proteins (in French). Les Technologies Separative par Membranes en Industries Agro-Alimentaires, Rennes, 10 May.

Berry, S. E. and M. H. Nguyen. 1988. High rate ultrafiltration of milk. Desalination 70: 169–76.

Bjorling, T and M. Olofsson. 1988. Cell debris filtration tests with Alfa-Laval microfiltration cartridge. Paper 5 read at the ACHEMA Annual Meeting, Sweden.

Boulton, R. 1984. The quantitative evaluation of juice and wine filtrations. Proc. Univ. of Calif. Davis Grape and Wine Symp. 16–19. Davis, CA.

Carre, Inc. 1986. Ultrapress apple juice process. Company Product Bulletin.

Castelas, B. and M. Serrano. 1990. Utilization of mineral membranes for the treatment of wine (in French). Proc. 1st Intl. Conf. Inorganic Membranes, 3–6 July, 283–90. Montpellier.

Cheryan, M. 1986. Ultrafiltration Handbook. p. 510. Technomic Publishing Co., Lancaster, PA.

Chun, U. H. and P. L. Rogers. 1988. Development of an ethanol/sorbitol process using a ceramic membrane cell recycle bioreactor. Desalination 70: 353–61.

Cueille, G. and M. Ferreira. 1990. Place of mineral membranes in the processes for bio-industry and the food-industry. Proc. 1st Intl. Conf. Inorganic Membranes, 3–6 July, 303–09. Montpellier.

Datar, R. 1984. Centrifugal and membrane filtration methods in biochemical separation. Filtration and Separation. 21(6): 402–06.

Daufin, G., J. P. Labbe, A. Quemerais, F. Michel and C. Fiaud. 1990. Fouling of Carbosep M5 membrane during ultrafiltration to clarify lactoserum (in French). *Proc. 1st Intl. Conf. Inorganic Membranes*, 3–6 July, 425–33. Montpellier.

Daufin, G., F. Michel and U. Merin. 1990. Performance of Carbosep M5 membrane in the micro- and ultrafiltration of lactoserum (in French). *Proc. 1st Intl. Conf. Inorganic Membranes*, 3–6 July, 483–86. Montpellier.

de la Garza, F. and R. Boulton. 1984. The modeling of wine filtrations. *Am. J. Enol. Vitic.* **35**(4): 189–94.

Filson, J. L., R. R. Bhave, J. Morgart and J. Graaskamp. 1991. Pyrogen Separations by Ceramic Ultrafiltration. U.S. Patent 5,047,155.

Finnigan, T., R. Shackleton and P. Skudder. 1987. Filtration of beer and recovery of extract from brewery tank bottoms using ceramic microfiltration. *Proc. Filtech Conf.* 22–24 September, 533. Utrecht.

Francoisse, O. and J. Hermia. 1989. Study of backflushing in cross-flow filtration using ceramic membrane. Final Report, Univ. Catholique de Louvain, Louvain la Neuve, Belgium.

Friedman, L. L., R. A. Hardwick, J. R. Danniels, R. R. Stromberg and A. A. Ciarkoeski. 1983. Evaluation of membranes for plasmapheresis. *Artif. Organs.* **7**: 435–42.

Galaj, S., A. Wicker, J. P. Dumas, J. Gillot and D. Garcera. 1984. Cross-flow microfiltration and backflushing on ceramic membranes. *Le Lait* **64**: 129–41.

Gillot, J., R. Soria and D. Garcera. 1990. Recent developments in the Membralox® ceramic membranes. *Proc. 1st Intl. Conf. Inorganic Membranes*, 3–6 July, 379–81. Montpellier.

Gillot, J. and D. Garcera. 1986. New ceramic filter media (in French). *Lait. Market.* **1007**: 23–30.

Goudedranche, H., J. L. Maubois, P. Ducruet and M. Mahaut. 1980. Utilization of the new mineral UF membranes for making semi-hard cheese. *Desalination* **35**: 243–58.

Grossiord, J. L. and G. Couarraze. 1983. *Rheology Initiation.* Ed. Techniques et Documentation, Paris.

Guibaud, J. 1990. Some applications of Membralox® ceramic membranes. *Proc. 1st Intl. Conf. Inorganic Membranes*, 3–6 July, 343–48. Montpellier.

Guibaud, J. 1991. (Personal communication).

Hanish, W. 1986. Cell Harvesting. In *Membrane Separations in Biotechnology, Bioprocess Technology*, ed. W. G. McGregor, pp. 61–68. Marcel Dekker, New York.

Harris, M. B. 1964. Grape juice clarification by filtration. *Am. J. Enol. Vitic.* **15**: 54–62.

van der Horst, H. C. 1990. Microfiltration in whey processing. *Proc. 1st Intl. Conf. Inorganic Membranes*, 3–6 July, 297–302. Montpellier.

Hoffmann, H., W. Kuhlmann, H. D. Meyer and K. Schugerl. 1985. High productivity ethanol fermentations with cross-flow membrane separation techniques for continuous cell recycling. *J. Membrane Sci.* **22**: 235–43.

Hoffmann, H., T. Scheper and K. Schugerl. 1987. Use of membranes to improve bioreactor performance. *Chem. Eng. J.* **34**: B13–B19.

Horiuchi, T., O. Otsubo, T. Uchima, R. Kusaba, H. Sugimoto, T. Yanagisawa, T. Inou, S. Nagata and S. Tanaka. 1981. Development of high performance membrane and optimum design of plasma separator. *Japan J. Artif. Organs* **10**: 304–07.

Ikeda, H. T., Tomono, A. Shimada, T. Hori, K. Ozawa, K. Sakai, H. Nakanishi and M. Inoue. 1986. In vitro evaluation of membrane plasma separator using human preserved blood. *Japan J. Artif. Organs* **13**: 914–17.

Kilham, O. W. 1987. Wine separation: Membranes, applications, Paper Read at the 5th Annual Membrane Technology/Planning Conference, Cambridge, MA.

Korolczuk, J., J. L. Maubois and J. Fauquant. 1987. Mechanization in the mold cheese industry (in French). *Milk—The Vital Force. 22nd Intl. Dairy Congress.* 123–28. D. Reidel Publishing Co., The Hague.

Kroner, K. H., H. Schutte, H. Hustedt and M. R. Kula. 1984. Cross-flow filtration in the downstream processing of enzymes. *Process Biochem.* **19**(2): 67–74.

Lozano, Y., O. Heich, M. Bennasar and B. Tarodo de la Fuente. 1986. Processing of pressed juice or purees of Kiwifruit by microfiltration with mineral membranes. IAA November, 1139–48.

Le, M. S. 1987. Recovery of beer from tank bottoms with membranes. *J. Chem. Tech. Biotechnol.* **37**(1): 59–66.

Mahaut, M., J. L. Maubois, A. Zink, R. Pannetier and R. Veyre. 1982. On the manufacture of cream cheese by membrane ultrafiltration of milk (in French). *Tech. Lait.* **961**: 9–13.

Mahaut, M., J. Korolczuk, R. Pannetier and J. L. Maubois. 1986. Fresh cream cheese making with ultrafiltration of acidified curd milk (in French). *Tech. Lait. Market.* **1011**: 24–28.

Malchesky, P. S., T. Horiuchi, J. J. Lewandowski and Y. Nose. 1989. Membrane plasma separation and the on-line treatment of plasma by membranes. *J. Membrane Sci.* **44**: 55–88.

Malmberg, R. and S. Holm. 1988. Low bacteria skim milk by microfiltration. *North. Eur. Food Dairy J.* **1**: 75–77.

Matsumoto, K., S. Katsuyama and H. Ohya. 1987. Separation of yeast by crossflow filtration with backwashing. *J. Ferment. Technol.* **65**(1): 77–83.

Matsumoto, K., M. Kawahara and H. Ohya. 1988. Crossflow filtration of yeast by microporous ceramic membrane with backwashing. *J. Ferment. Technol.* **66**(2): 199–205.

Maubois, J. L., G. Mocquot and L. Vassal. 1974. Milk and milk byproducts processing (in French). French Patent Appl. 7,436,025.

Maubois, J. L., A. Pierre, J. Fauquant and M. Piot. 1987. Industrial fractionation of main whey proteins. *IDF Bulletin* **212**: 154–59.

Maubois, J. L. 1990. (Personal communication).

Merin, U. and G. Daufin. 1990. Separation processes using inorganic membranes in the food industry. *Proc. 1st Intl. Conf. Inorganic Membranes*, 3–6 July, 271–81. Montpellier.

Meunier, J. P. 1990. Use of crossflow filtration for processing beer yeast and tank bottoms. *Proc. 5th World Filtration Congress*, 26–28. Nice.

Mietton-Peuchot, M. 1985. Tangential filtration in food industry (in French). *Ind. Alim. Agric.* **11**: 1185–88.

Mignonac, J. M., D. Poirier, P. Bourdin, M. Bennasar, B. Tarodo de la Fuente and J. Gillot. 1985. Filtration of wines and musts using mineral membranes (in French). *Proc. Filtra 84, Societe Francaise de Filtration*, 212–27. Paris.

Millipore Corp. 1974. Stabilization of wine. Publication AB 602, Bedford, MA.

Muir, D. D. and J. M. Banks. 1985. Developments in membrane technology. *J. Soc. Dairy Technol.* **38**(4): 116–19.

Nishizawa, Y., Y. Mitani, M. Tamai and S. Nagai. 1983. Ethanol production by cell recycling with hollow fibers. *J. Ferment. Technol.* **61**: 599–605.

Norton, Co. 1987. Ceraflo asymmetric ceramic microfilters. Company Product Bulletin.

Nose, Y., H. E. Kambic and S. Matsubara. 1983. Introduction to therapeutic apheresis. In *Plasmapheresis: Therapeutic Applications and New Techniques*, eds. Nose, Y. et al., P. S. Malchesky, J. W. Smith and R. S. Krakauer, pp. 1–22, Raven Press, New York.

Olesen, N. and F. Jensen. 1989. Microfiltration: The influence of operation parameters on the process. *Michwissenschaft* **44**(8): 476–79.

Ozawa, K., M. B. Kim, M. Sakurai, S. Takesawa and K. Sakai. 1985. Novel utilization of ceramic membranes in plasma treatment. In *Progress in Artificial Organs*, eds. Y. Nose, C. Kjellstrand and P. Ivanovich, ISAO Press, Cleveland.

Piot, M., J. C. Vachot, M. Veaux, J. L. Maubois and G. E. Brinkman. 1987. Skimming and producing low-bacteria milk by tangential micrifiltration (in French). *Tech. Lait. Market.* **1016**: 42–46.

Poirier, D., F. Maris, M. Bennasar, J. Gillot, D. Garcera and B. Tarodo de la Fuente. 1984. Clarification and stabilization of wines by tangential ultrafiltration using mineral membranes (in French). *Ind. Aliment. Agric.* **101**(6): 481–90.

Randerson, D. H. and J. A. Taylor, 1983. Protein adsorption and flux decay in membrane plasma separation. In *Plasmapheresis: New Trends in Therapeutic Applications*, eds. Y. Nose, P. S. Malchesky and J. W. Smith, pp. 69–80. ISAO Press, Cleveland.

Reed, R. J. R. 1986. Centenary review article. Beer filtration. *J. Inst. Brew.* **92**: 413–19.

Rios, G. M., B. Tarodo de la Fuente, M. Bennasar and C. Guizard. 1989. Crossflow filtration of biological fluids on inorganic membranes: a first state of the art. In *Developments in Food Preservation*, ed. S. Thorne, pp. 131–75. Elsevier Science Publishing Co., New York.

Ripperger, S. and G. Schulz. 1986. Microporous membranes in biotechnical application. *Bioprocess Eng.* **1**: 43–49.

Sakai, K., K. Ozawa, R. Mimura and H. Ohashi. 1987. Comparison of methods for characterizing microporous membranes for plasma separation. *J. Membrane Sci.* **32**: 3–17.

Sakai, K., K. Ozawa, K. Ohashi, R. Yoshida and H. Sakurai. 1989. Low-temperature plasma separation by crossflow filtration with microporous glass membranes. *Ind. Eng. Chem. Res.* **28**: 57–64.

Sandblom, R. M. 1978. Filtering process. U.S. Patent 4,105,547.

Shackleton, R. 1987. Filtration of beer using ceramic filters. U.K. Patent Appl. 2,176,715A.

Smith, J. W., Y. Asanuma, S. Suma, H. Harasaki, I. Zawicki, P. S. Malchesky and Y. Nose. 1979. Biocompatibility studies of hollow fiber plasma filtration for hepatic assistance. *Trans. Am. Soc. Artif. Int. Organs* **25**: 476–79.

Swientek, R. J. 1987. Metallic membrane filtration. *Food Processing*, **48**(1): 74–75.

Taddei, C. P. Aimar, G. Daufin and V. Sanchez. 1989. Role of some whey components on mass transfer in ultrafiltration. *Biotechnol. Bioeng.* **34**(2): 171–79.

Taniguchi, M., N. Kotani and T. Kobayashi. 1987. High-concentration cultivation of lactic acid bacteria in fermentor with crossflow filtration. *J. Ferment. Technol.* **65**: 179–84.

Tarodo, B. de la Fuente. 1990a. Rheological behavior of concentrates during the processing of lactic curds using inorganic membranes. University of Montpellier, Montpellier.

Tarodo, B. de la Fuente. 1990b. Clarification of strawberry and Kiwifruit purees using mineral membranes. University of Montpellier, Montpellier. (Unpublished work).

Tarodo, B. de la Fuente and C. Lecornu. 1988. Characterization of new ultrafiltration membranes; Parts I and II. University of Montpellier, Montpellier. (Unpublished work).

Thomas, R. L. 1988. The use of metallic membranes in fruit juice processing. Paper read at the 2nd Intl. Conf. of North American Membrane Society, Syracuse, 1–3 June.

Thomas, R. L., P. H. Westfall, Z. A. Louvieri and N. D. Ellis. 1986. Production of apple juice by single pass metallic membrane ultrafiltration. *J. Food Sci.* **51**(3): 559–63.

Tragardh, G. and P. E. Wahlgren. 1990. Removal of bacteria from beer using crossflow microfiltration. *Proc. 1st Intl. Conf. Inorganic Membranes*, 3–6 July, 291–95. Montpellier.

Venkataraman, K., P. K. Silverberg and M. T. Giles. 1988. Ceramic membrane applications in juice clarification—a case study. A paper presented at the 2nd Intl. Conf. of North American Membrane Society, Syracuse, 1–3 June.

Vetier, C., M. Bennasar and B. Tarodo de la Fuente. 1988. Study of the fouling of a mineral microfiltration membrane using scanning electron microscopy and physicochemical analyses in the processing of milk. *J. Dairy Research* **55**: 381–400.

Vetier, C., M. Bennasar and B. Tarodo de la Fuente. 1986. Milk mineral membrane interactions (in French). *Le Lait* **66**: 269–87.

Viikari, L. 1984a. Formation of levan and sorbitol from sucrose by *Zymomonas mobilis*. *Appl. Microbiol. Biotechnol.* **19**: 252–55.

Viikari, L. 1984b. Formation of sorbitol by *Zymomonas mobilis*. *Appl. Microbiol. Biotechnol.* **20**: 118–23.

Walla, G. and S. Donhauser. 1990. Filtration of beer and residual beer with crossflow microfiltration. *Proc. 5th World Filtration Congress*, 64–68 Nice.

9. Inorganic Membranes for the Filtration of Water, Wastewater Treatment and Process Industry Filtration Applications

R. R. BHAVE

Alcoa Separations Technology, Warrendale, PA

J. GUIBAUD

Societe des Ceramiques Techniques, Bazet

R. RUMEAU

University of Montpellier, Montpellier

9.1. INTRODUCTION

In the previous chapter on food and biotechnology applications the more prominent inorganic membrane applications were reviewed. The scope of inorganic membrane applications over the past 15–20 years has extended considerably beyond these and includes a wide variety of practical applications. Most of these can be classified into three major categories: (1) filtration of water, (2) wastewater treatment for environmental purposes or to recover valuable materials and (3) process industry filtrations.

This chapter reviews these applications, many of which have enjoyed some commercial success, while others are still under development. As will be evident, the published literature on the use of inorganic membranes is rather scanty. However, the specific and unique advantages of inorganic membranes are clearly demonstrated. Although a few technical challenges still remain, the barriers to commercialization are largely economic. With the growing concern for the preservation of the environment, the need for a superior performance device that meets or exceeds stringent regulations is becoming increasingly apparent. This challenge to perform well in harsh environments can be adequately met by the use of inorganic membranes.

9.2. FILTRATION OF WATER USING INORGANIC MEMBRANES

Inorganic-membrane-based filtration systems for the production of drinking water are used in commercial practice since 1984, primarily in Europe and more particularly in France. To date, there are about 10 installations in operation producing drinking-quality water in capacities varying from 7 m^3/h to 120 m^3/h (Duclert, Moulin and Rumeau 1990).

Figure 9.1 shows a typical industrial plant capable of producing up to 2000 m^3/day low-turbidity (0.1–0.2 NTU) bacteria-free drinking water. The

Figure 9.1. Industrial plant for the production of 2000 m^3/day of drinking water (plant built by OTV for the Sornin Syndicate (France) and operated by Compagnie Générale des Eaux).

filtration system is equipped with 120 m² of membrane surface area (Membralox® 0.2 μm Al_2O_3 membranes).

9.2.1. Cross-flow Microfiltration for the Production of Drinking Water

The production of drinking water by conventional technologies involves a number of sequential steps such as coagulation and disinfection with an oxidant (e.g., Cl_2, O_3), flocculation, clarification and final filtration. Media filters, particularly dual media filters (anthracite-sand, activated carbon-sand) are commonly used. The limitation of these filters is that these are suitable to filter low to moderate turbidity waters up to about 15–20 NTU (Bourdon, Bourbigot and Faivre 1988). For high-turbidity waters, their efficiency is quite low due to rapid fouling resulting in shorter filtration cycles to process and produce the required total throughput.

Cross-flow microfilters can perform the potabilization of karstic groundwaters in a single filtration step with minimal or no addition of chemical reagents. Bourdon, Bourbigot and Faivre (1988) have described in detail the performance of 0.2 μm pore diameter Al_2O_3 microfilters (Membralox®). The average physicochemical characteristics of the groundwater samples are given in Table 9.1. However, during certain periods (rainfall, melting snow, etc.) the turbidity level in karstic groundwater may increase very substantially up to about 100–120 NTU.

In order to simulate these upset conditions, synthetic water samples were prepared with the addition of turbid alluvial clay. Table 9.2 shows the raw-water and filtered-water (permeate) characteristics when filtration was performed using 0.2 μm Al_2O_3 membranes (Membralox®).

The cross-flow filtration of groundwater, like many other industrial process streams, is also limited by concentration polarization resulting on account of the accumulation of particles on the membrane surface during the

Table 9.1. Typical Physicochemical Characteristics of Groundwater Samples (Bourdon, Bourbigot and Faivre 1988)

Average Physicochemical Characteristics	Sample #1	Sample #2
Turbidity (NTU)	0.6–0.75	0.6–2
Organic matter (mg O_2/L)	0.3–0.6	0.1–0.7
Iron (μg/L)	40–60	10–70
Mn (μg/L)	≤ 20	≤ 20
Al (μg/L)	≤ 20	≤ 25–50

**Table 9.2.   Characteristics of Raw Water and Filtered Water
Using 0.2 μm Membralox® Microfiltration Membranes
(Bourdon, Bourbigot and Faivre 1988)**

Characteristics	Raw Water*	Permeate
Turbidity (NTU)	100–120	0.1–0.25
Organic matter (mg/O_2/L)	6.5–7.3	0.5
Al (μg/L)	2700	15
Fe (μg/L)	3000	40
Mn (μg/L)	45	> 4
Germs (20°C and 37°C)	> 1000	0

* Raw water is simulated karstic groundwater with the addition of
turbid alluvial clay.

course of filtration. Fouling is often minimized by the use of periodic
backflushing throughout the filtration cycle. Optimal operating conditions
are obtained by the evaluation of key process parameters such as cross-flow
velocity, ΔP_T, turbidity, backflushing and pretreatment.

Figure 9.2 illustrates the effect of cross-flow velocity on flux (Bourdon,
Bourbigot and Faivre 1988). An increase in the cross-flow velocity resulted in
higher flux. A cross-flow velocity of 5 m/s was judged to be optimal since
higher velocities provided only marginal flux increase at the expense of
substantially higher energy costs. Stable flux values of about 700 L/h-m^2 were
obtained at a ΔP_T of about 2.5 bar. The effect of turbidity on flux as a
function of time is shown in Figure 9.3. It is not clear if the flux decline
(approx. 30%) is entirely attributable to the increase in retentate water
turbidity from an initial value in the range 100–120 NTU to 750 NTU. The
ΔP_T was maintained at about 2.6 bar at a cross-flow velocity of 5 m/s.

In addition to turbidity reduction and bacteria removal, the production of
drinking water also requires the reduction of iron levels to acceptable limits.
The removal of iron from groundwater with cross-flow filtration using
inorganic and polymeric membranes was recently reported (Pain, Moulin
and Faivre 1990). Conventional treatment techniques involve the conversion
of divalent iron (Fe^{2+}) present in the dissolved state into the insoluble
trivalent form (Fe^{3+}) by atmospheric oxidation. The precipitated $Fe(OH)_3$
particles are then removed using media filters (e.g. sand filters).

Cross-flow micro- or ultrafiltration can remove iron hydroxide particles in
a single step. The characteristics of raw water and filtered water using
inorganic and polymeric membranes are shown in Table 9.3. These results,
obtained with Membralox® 50 nm ZrO_2 membrane and 3 nm nominal pore
size Romicon polysulfone membrane (e.g. deduced from molecular weight

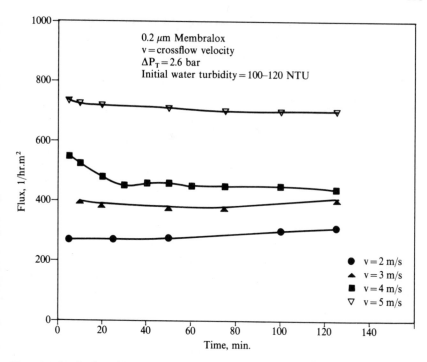

Figure 9.2. Production of drinking water using alumina membranes: effect of cross-flow velocity on flux (Bourdon, Bourbigot and Faivre 1988).

cutoff characteristics), indicate that the separation performance of the two UF membranes was similar despite the large difference in pore diameter values.

The water flux data for 50 nm and 0.2 μm ceramic membranes along with the 3 nm nominal pore size polymeric UF membrane are reported in Figure 9.4. Although the flux for the 0.2 μm membrane was higher than that for the smaller pore diameter membranes, it suffered from some irreversible fouling accompanied by a substantial flux decline. It was reported that for the 0.2 μm Al_2O_3 membrane, after 93 h of operation only 40% of the initial membrane permeability could be recovered. For the 50 nm ZrO_2 membrane, even after 245 h of operation 75% of the initial water permeability was obtained after chemical regeneration of fouled membranes. High flux recovery after over 200 h of operation was also observed for the 3 nm polysulfone membrane (Pain, Moulin and Faivre 1990). These results are considered preliminary. Additional work is required to establish optimal operating conditions including the choice of a proper membrane pore diameter and cleaning procedure (e.g. by O_3).

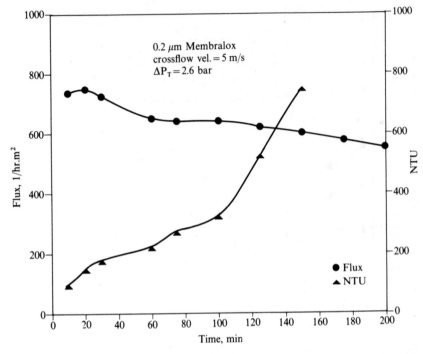

Figure 9.3. Effect of increasing water turbidity values on flux as a function of time (Bourdon, Bourbigot and Faivre 1988).

Table 9.3. Performance of Inorganic and Polymeric UF Membranes in the Filtration of Groundwater (Pain, Moulin and Faivre 1990)

Parameters	Units	Raw Water	Filtered Water	
			Polymeric Membrane (3 nm)	Inorganic Membrane (50 nm)
Turbidity	NTU	0.50	0.10	0.10
Fe	μg/L	300	Trace*	Trace
Mn	μg/L	25	25	25
Aerobic germs (24 h, 37°C)	nb/ml	200	0	0
Aerobic germs (72 h, 20°C)	nb/ml	500	0	0

* Below detectable limit

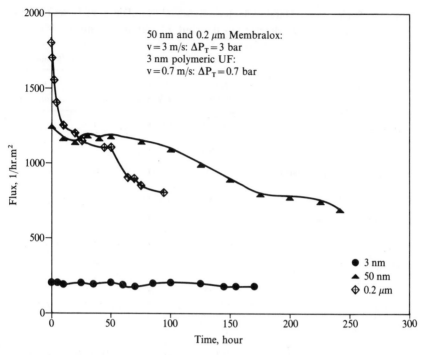

Figure 9.4. Effect of pore diameter on microfiltration of groundwater (Pain et al. 1990).

9.2.2. Bacteria Removal from Water by Inorganic Membrane Filters

The filtration of water to produce essentially bacteria-free water has been practiced on the industrial scale for the past several years using inorganic membrane filters (Duclert, Moulin and Rumeau 1990). The objective of the filtration process is typically not limited to bacteria removal alone but also to the reduction of water turbidity to low values (0.1–0.2 NTU) and to reduce the concentration of certain elements (e.g. iron, aluminum and manganese) to within acceptable limits. Some of these aspects are reviewed in the preceding Section 9.2.1.

Several different water sources such as groundwater, river water and lagoon water may contain large concentrations of bacteria. The performance of 0.2 μm Al_2O_3 membranes for bacteria removal was recently described (Duclert, Moulin and Rumeau 1990). Table 9.4 summarizes the results for bacteria removal from various contaminated water streams. The proper selection of pore diameter is very critical to ensure that colloids, if present, will not penetrate into the membrane structure to any significant extent.

Table 9.4. Performance Data on Bacteria Removal from Different Types of Water Using 0.2 μm Al_2O_3 Membranes (Duclert, Moulin and Rumeau 1990)

	Before Filtration		After Filtration	
	Total Coliforms	Fecal Coliforms	Total Coliforms	Fecal Coliforms
Lagooning outlet	20,000 +/− 10,000	1,500 +/− 500	0	0
Entry to activated slurry station	80,000 +/− 300,000	40,000 +/− 12,000	0	0
Exit from activated slurry station	45,000 +/− 25,000	1,100 +/− 200	0	0
River water	800 +/− 200	100 +/− 50	0	0

Membrane processes are susceptible to rapid irreversible fouling in the presence of colloids.

The excellent performance of 0.2 μm Al_2O_3 membranes for the complete removal of total coliforms (including fecal coliforms) from their high feed levels up to 300,000 is indeed noteworthy. This study also shows the reduction of turbidity in raw water ranging from 5–20 NTU to values below 0.5 NTU in the filtered water (Duclert, Moulin and Rumeau 1990).

9.3. NON-OILY WASTEWATER TREATMENT WITH INORGANIC MEMBRANES

There are numerous sources of non-oily wastewaters that are amenable to filtration using inorganic membranes. Some of the more important of these discussed in this section are based on published literature. Although such published data are rather scanty, many applications are believed to be under active evaluations in many parts of the world.

9.3.1. Treatment of Textile and Paper Industry Effluents

A few studies on the use of inorganic membranes for the treatment of textile and paper industry effluents are reported in the literature (Brandon 1984, Jonsson and Petersson 1988; Soma, Rumeau and Sergent 1990). Many of these studies deal with the removal of dyes and organics from effluents to meet environmental discharge restrictions. Although technically feasible, these applications have to date enjoyed limited success, primarily due to economic considerations and lack of a rigorous implementation of discharge regulations. This situation is likely to change in future due to the emerging trend to impose tighter environmental standards. When this happens, the somewhat higher capital cost requirement of inorganic membrane processes as compared to conventional treatment techniques (e.g. chemical precipitation, biological treatment, etc.) may not remain an important barrier.

Table 9.5 summarizes the membrane performance data on the removal of dyes and organics from textile and paper industry effluents. Most of these processes are still under development. The most notable amongst these is the removal of certain direct dyes used in acrylic fiber manufacture (Brandon 1984). In this process, dynamic $Zr(OH)_4$–polyacrylic acid (PAA) reverse-osmosis-type membranes on porous tubular sintered stainless steel supports were used. Although the technical feasibility of the process was clearly demonstrated (at a textile plant in the U.S.), it was judged to be not very cost effective. The value of recovered dyes and cost savings in water and wastewater treatment were not sufficient to justify the commercial viability of the process (Brandon 1984).

An interesting study was reported (Soma, Rumeau and Sergent, 1990) concerning the removal of soluble and insoluble dyes with inorganic micro-filtration membranes (Membralox® 0.2 μm Al_2O_3). It is evident from Table 9.5 that microfiltration membranes are very effective in removing insoluble dyes but require the use of surfactants to improve removal efficiency of soluble dyes. By the addition of surfactants such as sodium dodecyl sulfonate, cetyl trimethyl ammonium bromide, Lavotan NN containing anionic surfactants and Ceranine HCS (a cationic fatty acid derivative) the dye removal efficiency was dramatically improved. On the other hand, to remove low molecular weight organics such as yellow acid (MW = 749) a pore diameter of about 1 nm is required (Guizard, Larbot and Cot 1990).

The use of ceramic membranes to treat effluents from a conventional bleach plant for Kraft pulp is also described (Jonsson and Petersson 1988). For the 0.2 μm Al_2O_3 membrane, the flux values ranged from 150 L/h-m² to 1300 L/h-m² for the various streams tested. The retention of organics (as measured from the reduction in COD values) was quite low ranging from only 25%–45%. It appears that low molecular weight cutoff UF membranes (below MW 2000) will be required to obtain acceptable levels of COD reduction.

Nevertheless indigo dye recycling is now performed industrially using 0.2 μm Membralox® Al_2O_3 membrane in a textile dyeing process line (Guibaud 1991).

Table 9.5. Performance Data on the Removal of Dyes and Organics From Textile and Paper Industry Effluents Using Inorganic Membranes

Type of Dye and/or Chemical	%Dye/ Chemical In Retentate	Membrane Identification	Membrane Pore Diameter (nm)	Temperature Range (°C)	Transmembrane Pressure (bar)	% Color Rejection by Membrane	Average Flux (L/h·m²)	Reference
Direct dye	1–2	$Zr(OH)_4$–PAA (Carre)	—	60–80	3.5–5	86–99	100–150	Brandon (1984)
Yellow acid (MW = 749)	0.1	Heteropolysiloxane	~1	20	2	100	<10	Guizard, Larbot and Cot (1990)
Soluble dyes without surfactant	0.5–2	α-Al_2O_3 (Membralox®)	200	20	3	14–62	20–40	Soma, Rumeau and Sergent (1990)
Soluble dyes with surfactant	0.5–2	α-Al_2O_3 (Membralox®)	200	20	3	96–98	60–100	
Insoluble dyes	0.5–1	α-Al_2O_3 (Membralox®)	200	20	3	98	100–200	Soma, Rumeau and Sergent (1990)

9.3.2. Concentration of Latex Wastewaters

The use of membranes for the concentration of latex wastewater was first successfully demonstrated in the mid-1970s (Mir, Eykamp and Goldsmith 1977, Bansal 1976a). Despite the early commercial success, there appears to be very few published reports on the use of inorganic membranes for this application in the 1980s.

The concentration and/or recovery of latex from wastewaters is conventionally performed using evaporation technology. The concentration of dilute latex wastewater from about 0.5% to 25% is largely performed to reduce the volume of wastewater for the environmental pollution control purposes. The latex can be further concentrated by evaporative processes or by chemical coagulation. Certain latex waters (e.g. polyvinyl chloride) are concentrated from 30% to levels between 40% and 65% to recover the latex.

The important process parameters in this application are cross-flow velocity, ΔP_T and latex concentration (Mir, Eykamp and Goldsmith 1977, Bansal 1976a). Figures 9.5 and 9.6 illustrate the performance of inorganic and

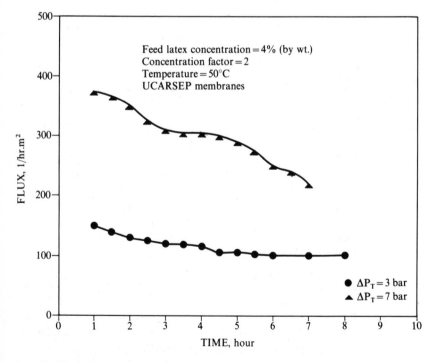

Figure 9.5. Performance of dynamic ZrO_2 membranes in the concentration of latex (Bansal 1976a).

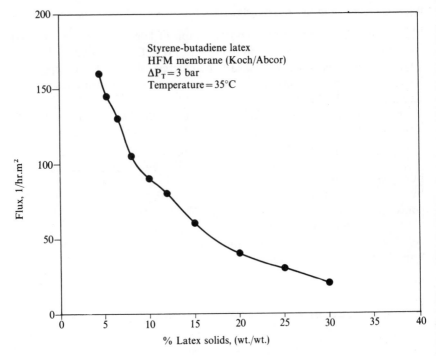

Figure 9.6. Performance of polymeric UF membranes in the concentration of latex waters (Mir, Eykamp and Goldsmith 1977).

polymeric membranes, respectively, for the filtration and concentration of latex wastewater. These data clearly show the superior performance of the inorganic membrane relative to that of the polymeric UF membrane. The membrane used in the studies reported by Bansal (1976a) consists of a dynamically formed ZrO_2 membrane on porous carbon supports (Ucarsep, Gaston County).

For the Ucarsep membrane, the average flux for 4% latex solids at a ΔP_T of approximately 3 bar was 150 L/h-m² as compared to 100 L/h-m² obtained for the HFM noncellulosic membrane (Koch/Abcor). With inorganic membranes, higher ΔP_T can be employed resulting in substantially higher flux. Polymeric membranes are capable of operating over a limited ΔP_T as they suffer from compaction at higher operating pressures. The effect of ΔP_T on flux is also illustrated in Figure 9.5. At a ΔP_T of 7 bar, average flux values of 225 L/h-m² were obtained. This value is about 50% higher than the value at a ΔP_T of about 3 bar. The favorable effect of temperature on flux was also demonstrated, but higher temperatures (> 50°C) result in agglomeration of latex particles which adversely affects the flux stability. Latex concentration is therefore typically performed at temperatures between 20 and 35°C.

9.3.3. Miscellaneous Effluent Treatment Applications

Inorganic membranes have also been evaluated in other effluent treatment applications. These performance descriptions indicate the developmental nature of these applications. No commercial installations are reported to be in operation at the present time.

Treatment of Sewage Water Al_2O_3 UF membranes were evaluated on the pilot-scale to remove bacteria from sewage water. γ-Al_2O_3 Membralox® membranes with a pore diameter of 6 nm were used. The filtrate was not only bacteria free but also low in nitrates, sulfates and oxyanions relative to the high concentrations (> 100 ppm) encountered in sewage samples. A flux value of 100 L/h-m² was sustained over a long period of operation (1000 h) at a ΔP_T of 20 bar (Guizard et al. 1984). The results are summarized in Table 9.6. Zirconia UF membranes with a similar pore diameter are under evaluation because of their higher (e.g. up to 10 times) permeability compared to γ-Al_2O_3 membranes (Guibaud 1991).

Treatment of Food Processing Effluents Inorganic membranes were also evaluated for the treatment of cottonseed protein and soy-flour protein extracts. Dynamic ZrO_2 membranes on porous carbon supports (MW cutoff = 20,000) were used for this application (Bansal 1976b).

In the case of cottonseed protein extract, the feed was concentrated from 1.6% solids (wt./wt.) to 14.6% solids. An average flux of 150 L/h-m² was

Table 9.6. Use of Inorganic UF Membranes in the Treatment of Food Processing and Sewage Water Effluents

Type of Effluent	Membrane Identification	Pore Diameter (nm)	Temperature Range (°C)	Trans-membrane Pressure (bar)	Flux (L/h-m²)	Reference
Cottonseed (protein exract)	ZrO_2 on carbon UCARSEP	*	65	5	150	Bansal (1976b)
Soy-flour (protein extract)	ZrO_2 on carbon UCARSEP	*	65	5	135	Bansal (1976b)
Sewage water	γ-Al_2O_3 (Membralox®)	6	20	20	100[†]	Guizard et al. (1984)

* Not specified, 20,000 MW cutoff
† Sustained over 1000 h of operation; filtrate was bacteria free

obtained. The average flux for the concentration of soy-flour protein was about 135 L/h-m^2. The effluent stream was concentrated from 4.8% solids to 13.1% total solids.

9.4. OILY WASTEWATER TREATMENT WITH INORGANIC MEMBRANES

This application area was vigorously pursued in the early 1970s with the advent of dynamic ZrO_2 membranes on porous carbon tubes (Ucarsep) first developed by Union Carbide. Although inorganic membranes enjoyed some commercial success, they appear to have lost the market share to more favorable economics with conventional technologies and polymeric membranes. In recent years, inorganic membranes are gaining attention in new applications such as treatment of oily produced water and recovery of machining coolants.

9.4.1. Cross-flow Filtration of Oily Produced Water

In recent years, there is a growing interest in the removal of oily substances and suspended solids from produced water using ceramic membranes (Chen et al. 1989, Chen et al. 1991). It is necessary to treat produced water to meet the specifications set by regulatory agencies such as the U.S. EPA (United States Environmental Protection Agency). Treated water that meets the standards for oil, grease and suspended solids may be considered safe for overboard discharge from offshore facilities or may be used in reinjection for waterflood.

Chen et al. (1991) have described the use of ceramic membranes in the cross-flow filtration of produced water at several onshore and offshore locations. Typical produced water characteristics from offshore facilities (Gulf of Mexico) are given in Table 9.7. Ceramic microfiltration membranes of pore diameter in the range 0.2–0.8 μm were evaluated in these developmental studies. Produced water samples were chemically pretreated to produce a floc of discrete solids to minimize fouling due to particle penetration in the membrane structure.

In the absence of chemical pretreatment rapid fouling was observed. This suggests that in situations where chemical pretreatment is not a viable option small pore diameter UF membrane may be required. The flux for the cross-flow filtration of oily wastewater using inorganic UF membranes is typically lower than that reported with microfiltration membranes (Bhave and Fleming 1988).

The concentration of oily substances in the filtrate (permeate) was at a very low level (typically 2–5 ppm). The concentration of oily substances in the feed ranged from 100–575 ppm. Total suspended solids were reduced from 100–390 ppm to below 1 ppm in the filtrate. Oily substances were concen-

Table 9.7. Typical Produced Water Analysis
(Chen et al. 1991)

Parameter	Sample # A*	Sample # B*
Na^+ (mg/L)	76,300	52,248
Ca^{2+} (mg/L)	3,850	3,840
Mg^{2+} (mg/L)	1,900	2,256
Fe^{3+} (mg/L)	25–40	57
Ba^{2+} (mg/L)	108	50
Cl^- (mg/L)	130,000	94,000
HCO_3^- (mg/L)	87	204
SO_4^{2-} (mg/L)	0	5
CO_3^{2-} (mg/L)	0	0
OH^- (mg/L)	0	0
H_2S (mg/L)	< 0.1	< 0.1
TDS (mg/L)	212,100	152,660
TSS (mg/L)	100–290	73–350
pH	6.0	7.0
Temperature (°C)	32–38	32–40
Oil and grease (mg/L)	28–108	105–574

* Samples A and B are obtained from two different locations in the Gulf of Mexico

trated in the retentate up to a level of only about 1%. In this application, the primary objective is the treatment of produced water to render it safe for discharge. The concentration and/or recovery of oil is often secondary as long as the oil-rich retentate is reduced to a few percent of the volume of treated produced water.

A proper selection of pore diameter is critical not only to maximize throughput but also to ensure that the discharge standards for oil and suspended solids are consistently met. For the economic viability of the cross-flow filtration process, fouling must be minimized and complete regeneration of fouled membranes demonstrated to assure long operational reliability (3 years or more) before membrane replacement becomes necessary.

Figure 9.7 shows the effect of pore diameter on ΔP_T as a function of time to maintain a constant flux of about 1500 L/h-m². It was observed that the 0.2 μm pore diameter Membralox® Al_2O_3 membrane showed a considerable increase in ΔP_T over time as compared to the 0.5 μm membrane. Periodic backflushing was used to prevent excessive accumulation of foulants on the membrane surface and to achieve high flux throughout the filtration operation.

The effectiveness of backflushing depends on the type of fouling observed for a given application. Backflushing is quite effective if the flux is largely

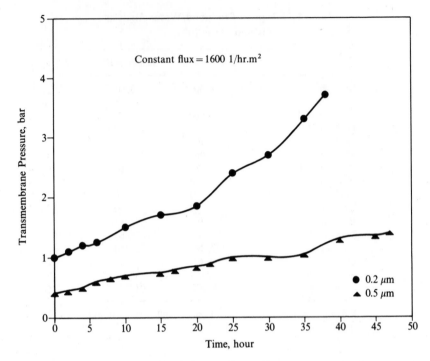

Figure 9.7. Produced water filtration: influence of pore diameter on transmembrane pressure increase (fouling) (Chen et al. 1991).

controlled by polarization due to accumulation of particles on the membrane surface. However, if fouling is caused by chemical effects such as adsorption of chemical substances or penetration of colloidal materials in the membrane structure, backflushing is not very effective resulting in rapid irreversible membrane fouling. For the filtration of chemically pretreated produced water the effectiveness of backflushing suggests that the filtration is controlled by polarization effects.

The performance of the 0.8 μm membrane was found to be similar to the 0.5 μm membrane although its operational stability over longer periods of operation was somewhat poor. These results point to 0.5 μm as the optimal pore diameter membrane to obtain relatively high flux with operational stability.

9.4.2. Treatment and/or Recovery of Oils from Oily Wastes and Oil–Water Emulsions

A number of membrane filtration processes applied to the treatment and/or recovery of oils from oil–water emulsions are described (Bansal 1975, 1976a,

Table 9.8. Performance Data of Inorganic UF Membranes in the Treatment and/or Recovery of Oils from Oily Wastes

Type of Oily Wastewater	Membrane/ Pore Diameter	Average Oil Concentration		Average Transmembrane Pressure (bar)	Average Temperature (°C)	Average Flux (L/h·m²)	Oil Rejection (%)	Reference
		Retentate (%)	Permeate (ppm)					
Lubricating oil	UCARSEP* (−) Membralox (4–50 nm)	12	50	5	65	85	99.64	Bansal (1975)
		0.1	5	3	50	75	99.5	Bhave and Fleming (1988)
Cutting oil	UCARSEP	12	15	5	65	120	99.8	Bansal (1975)
Drawing oil	UCARSEP	12	500	5	60	90	99.5	Bansal (1976a)
	UCARSEP	25	1300	3	57	40	99.5	Bansal (1976a)
Rolling oil	UCARSEP	25	40	5	65	75	99.99	Bansal (1975)
Vegetable oil	UCARSEP	15	1000	5	65	70	99.5	Bansal (1975)
	Membralox® (0.2 μm)	0.3	<10	1	60	27	99.7	Bhave and Fleming (1988)
		0.3†	<10	0.5	60	550	99.7	Bhave and Fleming (1988)

* Not specified, about 20,000 MW cutoff
† With chemical pretreatment

Bhave and Fleming 1988, Falletti 1988, Guibaud 1989). These largely deal with oily wastes generated in cutting, grinding and rolling operations. The treatment of waste lubricating and vegetable oils, and oil–water emulsions containing machining coolants is also described. The treatment temperatures vary from ambient up to 65°C with pH values ranging from 5–14.

Treatment and/or Recovery of Oils from Oily Wastes In this category, a number of different waste oils were evaluated to determine the techno–economic viability of inorganic membrane filtration systems. Some results are summarized in Table 9.8. Amongst the first to be evaluated were the dynamic ZrO_2 membranes on porous carbon supports (Ucarsep). Since 1973, a number of filtration systems were installed in automotive and other commercial plants involving extensive machining operations (Banal 1975).

In the early 1980s, Tech Sep (formerly SFEC) developed ZrO_2 membrane layers integrally bonded onto porous carbon tubes. There has been, however, very little published work on the use of these composite membranes, perhaps due to proprietory considerations.

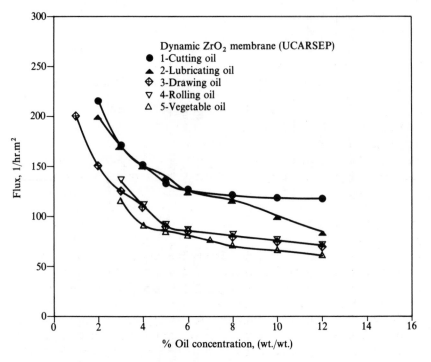

Figure 9.8. Cross-flow filtration of oily wastes: flux versus oil concentration (Bansal 1975).

Inorganic membranes show very high oil rejection characteristics. Consequently, the permeate oil concentrations are quite low (typically below 50 ppm) and often below 10 ppm with 50 nm Membralox® UF membranes. They also show an ability to concentrate waste oils from low concentrations (typically a few hundred ppm) to concentrations of 25% and higher. Oil concentrations above 25% are obtainable but are likely to affect the process economics adversely. The effect of oil concentration on flux is illustrated in Figure 9.8. As may be expected, an increase in retentate oil concentration was accompanied by a substantial flux decline (Bansal 1975).

At retentate oil concentrations above 10% a significant increase in permeate oil concentration was observed. The overall oil rejection efficiency was maintained at 99.5% or better at retentate concentrations up to 25% (wt./wt.). The ultrafiltration concentrate can be further concentrated in the range 50–70% by conventional chemical treatment or concentration techniques.

Polymeric membranes are also described for the concentration of oil–water emulsions (Mir, Eykamp and Goldsmith 1977). Flux values in the range 45–70 L/h-m^2 are reported for noncellulosic solvent and pH-resistant membranes (Koch HFM membranes). Membralox® zirconia membranes with 50 nm pore diameter typically show a flux of about 200 L/h-m^2 with permeate oil concentration lower than 10 ppm (Guibaud 1991). The performance of inorganic membranes can not only be superior to their polymeric counterparts, but their ability to withstand harsher environments allow the concentration of oil to higher levels with longer operating life (2 years or more). These operational characteristics may very well be advantageous in many situations.

Dynamic Zr(OH)$_4$–PAA reverse-osmosis-type membranes are also described for the treatment of emulsified oily wastewater from a metal-parts washing effluent (Spencer 1990). A low flux of only 20–25 L/h-m^2 was obtained. The process was operated at about 50% recovery with concentrate recycle. A high overall volume reduction of about 96% was achieved.

Recovery and Reuse of Machining Coolants A particularly interesting emerging application area for ceramic membranes is the recovery and reuse of machining coolants. Ceramic membranes have demonstrated an excellent ability to remove particles and oil/grease contaminants while preventing loss of valuable coolants which are essentially oil–water emulsions (Falletti 1988, Guibaud 1989). Table 9.9 summarizes the results for the treatment of machining coolants using Membralox® ceramic membranes.

The proper selection of membrane pore diameter and material properties of the membrane layer in contact with the contaminated coolant is very critical. In addition, optimization of cross-flow velocity and ΔP_T is also important to maximize flux without sacrificing permeate quality. Table 9.9

Table 9.9. Performance Characteristics of Inorganic Membranes for the Recovery and Recycle of Machining Coolants/Cleaners

Sample Identification	Membrane Tested	Pore Diameter (μm)	Flux (L/h-m²)	Quality of Permeate	Comments
Water-based metal working lubricant (Cimcool MB602*)	Membralox® Alumina	3	100	Particle free; breakage of oil–water emulsion	Probably unsuitable
	Modified A Membralox®	5	2200	Particle free with no loss of coolant	Difficult to regenerate
	Modified B Membralox®	5	1500	Particle free with no loss of coolant	Longer term flux decline
	Modified B Membralox®	5	2500	Particle free with no loss of coolant	Stable flux
Spent cleanex solutions			700	Free of oil; cleaner solution quality	Stable flux after several regenerations
Synthetic machining coolants mixture (Parcobound 60†)	Membralox® Zirconia	0.05	160	Very clear, no oil in permeate	Minimal fouling and easy to clean compared to 0.1 μm and 0.2 μm Membralox®

* Cincinatti Milacron
† Parker Industries

also illustrates the importance of material surface properties in the recovery of water-based metal-working lubricants commonly used in the auto industry.

For the 3 μm Membralox® alumina membrane, coalescence of oil–water emulsion was observed with the consequent plugging of the pores by oil droplets. This resulted in a low flux of about 100 L/h-m^2. On the other hand, no emulsion breakage was observed for the 5 μm Membralox® Al_2O_3 membrane. There was minimal fouling, and a high flux of about 2200 L/h-m^2 was sustained. The fouled membranes were however difficult to regenerate with hot cleaning solutions indicating that strong chemical fouling may have occurred.

Modified Membralox® membranes (Type B) gave stable performances with minimal fouling. These modifications are believed to change the surface properties of the membrane layer and render them more resistant to surface fouling.

In the case of synthetic machining coolants, Membralox® UF membranes with 50 nm pore diameter were reported to be highly efficient for the removal of particles and oil from contaminated coolants (Guibaud 1989). Although the flux values for the regeneration of water-based lubricants were lower than those observed with the larger pore membranes, the process was very stable. Fouled membranes were relatively easy to regenerate as compared to microfiltration membranes. This suggests that with UF membranes there is no particle penetration in the membrane layer. This aspect may be critical in some applications where chemical fouling of ceramic membranes cannot be eliminated through the modification of surface properties.

9.5. INORGANIC MEMBRANES IN PROCESS INDUSTRY FILTRATION

Inorganic membranes, due to their high chemical, thermal and mechanical resistance, are well suited for applications involving harsh environments where polymer membranes are unsuitable and conventional technologies are either inadequate or very energy intensive. In many process industry applications, inorganic membranes provide a technologically acceptable alternative to handling difficult processing problems. Despite these potential advantages, inorganic membranes have not attained wide commercial acceptance.

There are only a few descriptions on the use of inorganic membranes in process industry applications other than those specific applications covered in the preceding sections. Purification of concentrated $CdSO_4$ solutions containing low amount of $FeSO_4$ by oxidation of Fe^{2+} to Fe^{3+}, precipitation and microfiltration using ceramic membranes has been studied but no data are reported (Rumeau 1991). In addition, Membralox® ceramic membranes (e.g. 0.8 μm pore diameter) are believed to be in use in a significant number of small installations for the recycling of alkaline cleaners (Guibaud 1991). These plants typically operate at 40–60°C at low transmembrane pressure with backflush. Flux values range between 300–600 L/h-m^2 for 1–3 g/L oil/grease concentrations in the retentate.

Two areas that have received some attention are the recovery of caustic from process effluents and hydrocarbon processing applications. These are described in the following sections.

9.5.1. Recovery of Caustic from Process Effluents

Caustic soda (NaOH) is widely used in the textile industry to clean fabrics in preparation for dyeing. Caustic is expensive and the discharge of dirty caustic waste is a serious environmental problem. Conventionally, dirty caustic is neutralized by acid addition and discharged to a municipal waste treatment plant.

Dynamic reverse-osmosis-type membranes of $Zr(OH)_4$–PAA on porous tubular sintered stainless steel supports (Carre, a subsidiary of Du Pont Separations) are described for caustic discharge reduction in textile processing (Brandon 1986, Fulmer 1985, Gaddis, Spencer and Jernigan 1989). These membrane systems have been in commercial use since 1986, with capacities up to 800 m^3/day.

The membrane filtration process recovers about 80–90% of NaOH which is recycled and reused. The unrecovered caustic (typically 10–20%) is sent to effluent treatment facilities. Figure 9.9 shows the schematic for NaOH recovery using dynamic $Zr(OH)_4$–PAA membrane systems. Caustic recycle processes operate at a temperature of about 95°C, at a ΔP_T of up to 60 bar. The concentration of NaOH in process effluents typically ranges from 3–10% (by wt.). Average flux values of about 20 L/h-m^2 are realized (Fulmer 1985). The system is cleaned daily and the dynamic membranes are reformed weekly.

9.5.2. Hydrocarbon Processing Applications

Ceramic membranes, due to their exceptional temperature resistance, are ideally suited for high-temperature applications where polymeric membranes are unsuitable. Their high mechanical strength and resistance to abrasion may also be advantageous in situations where high solids concentrations and pressures are encountered. At the present time, there are no known commercial processes using ceramic membranes but a number of these are believed to be in various stages of development.

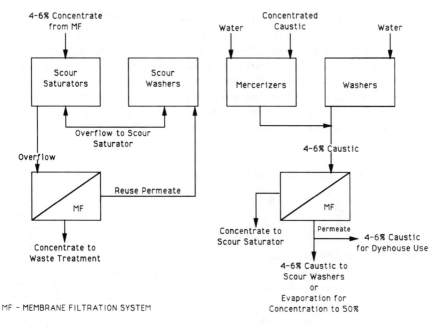

Figure 9.9. Process schematic for caustic recovery and recycle using dynamically formed inorganic membranes (Carre Inc., 1985).

High-Temperature Ultrafiltration for the Refining of Waste Oils   Inorganic membranes may be advantageously utilized for the regeneration of waste oils such as spent motor oils. Although this application has not yet been fully commercialized, a pilot plant with a capacity to regenerate 5000 tons/yr of spent motor oils (Total- CEA Regelub process) and equipped with inorganic membranes (Carbosep® UF membranes) capable of operating up to 300°C is in operation (Charpin and Rigny 1990).

Waste oils may be generated as a result of auto engine oil change and may contain up to 20% impurities (Arod et al. 1984, Biederman and Rossarie 1984). Some of the impurities are identified as water, sludge, carbonaceous particles, metallic particles and machining liquids. Conventional processes for the regeneration of waste oils use acid treatment following centrifugation to flocculate the carbon particles and sulfonate the oxidized products (Arod et al. 1984). The disposal of acid-sludge is a growing environmental concern. Additionally, new high-performance lubricating oils contain large amounts of additives requiring increasing quantities of acid and clay further aggravating the disposal problem.

Mineral UF membranes based on metal oxides (such as Al_2O_3, MgO and SiO_2) with pore diameters in the range 20–40 nm were used to refine waste motor oils. The process was operated at temperatures from 200–350°C at a ΔP_T of 5 bar and cross-flow velocity in the range 3.2–5 m/s. Flux values ranged from 20–30 L/h-m^2 for a volumetric concentration factor of 3. Periodic backflushing was used at pressures up to 30 bar.

Separation of Catalyst from Hydrocarbon Residues In hydrocarbon conversion applications, cross-flow micro- or ultrafiltration may be used to remove catalyst particles from converted oil. Membralox® ceramic membranes with pore diameters ranging from 15 nm–0.3 μm were evaluated at temperatures up to 250°C and a ΔP_T of 5 bar. For 15 nm membranes stable flux with catalyst rejection of > 98% was achieved. Higher pore diameter membranes showed only about 70% particle rejection and substantial flux decline due to concentration polarization or pore blockage by particles (Deschamps et al. 1990).

The flux was also found to be a strong function of viscosity of the hydrocarbon residue. It was observed that at a fixed catalyst concentration, an increase in temperature caused a larger (corresponding to the decrease in viscosity) than expected increase in flux. For instance, at 100°C for a catalyst particle suspension with a viscosity of 2.9 cP, the permeate flux was 54 L/h-m^2. The ΔP_T was 5 bar and cross-flow velocity of 4 m/s. At 145°C and a viscosity of 2.3 cP a substantially larger flux of 92 L/h-m^2 was obtained, under otherwise uniform conditions.

This phenomenon may be explained by the formation of a dynamic membrane resulting from the deposition of high molecular weight products, the thickness of which decreases rapidly with increasing temperatures. This application illustrates the advantage of using ceramic membranes for the high-temperature filtration of catalyst from hydrocarbon residues. Additional work is needed to further characterize, develop and optimize such hydrocarbon processing applications in order to bring them to commercial fruition.

REFERENCES

Arod, J., B. Bartoli, P. Bergez, J. M. Biedermann, P. Caminade, J. M. Martinet, J. Maurin and J. Rossarie. 1984. Process for the treatment of a hydrocarbon charge by high temperature ultrafiltration. Canadian Patent 1,168,590.

Bhave, R. R. and H. L. Fleming. 1988. Removal of oily contaminants in wastewater with microporous alumina membranes. *A.I.Ch.E. Symp. Ser.* 84(261): 19–27.

Bansal, I. K. 1975. Ultrafiltration of oily wastes from process industries. *A.I.Ch.E. Symp. Ser.* 71(151): 93–99.

Bansal, I. K. 1976a. Concentration of oily and latex wastewaters using ultrafiltration inorganic membranes. *Ind. Water Engr.* 13(5): 6–11.

Bansal, I. K. 1976b. Progress in developing membrane systems for treatment of forest products and food processing effluents. *A.I.Ch.E. Symp. Ser.* 73(166): 144–51.

Biederman, J. M. and J. H. Rossarie. 1984. Ultrafiltration improves re-refining of waste oils. CEA, France, (unpublished work).

Bourdon, F., M. M. Bourbigot and M. Faivre. 1988. Crossflow microfiltration of groundwaters of karstic origin (in French). *L'Eau Industrie les Nuisances* 121, 35–41.

Brandon, C. A. 1984. Closed cycle textile dyeing: Extended evaluation of full-scale hyperfiltration demonstration (project summary). U.S. EPA 600/S2-84-147.

Brandon, C. A. 1986. Hyperfiltration for textile discharge reduction (project summary). U.S. EPA 600/S2-85/139.

Charpin, J. and P. Rigny. 1990. Inorganic membranes for separative techniques: from uranium isotope separation to non-nuclear fields. *Proc. 1st Intl. Conf. Inorganic Membranes*, 3–6 July, 1–16. Montpellier.

Chen, S. C., J. T. Flynn, R. G. Cook and A. L. Casaday. 1989. Ceramic crossflow microfiltration of produced water. Paper read at the American Filtration Society Regional Meeting, 30 October–1 November, Houston.

Chen, S. C., J. T. Flynn, R. G. Cook, and A. L. Casaday. 1991. Removal of oil, grease and suspended solids from produced water using ceramic crossflow microfiltration. *SPE Production Engineering* (in press).

Deschamps, A., C. Walther, P. Bergez and J. Charpin. 1990. Application of inorganic membranes in refining processes of petroleum residues. *Proc. 1st Intl. Conf. Inorganic Membranes*, 3–6 July, 237–42. Montpellier.

• Duclert, F., C. Moulin and M. Rumeau. 1990. Results of six years of practical experience in the potabilization of water by ultrafiltration. *Proc. 5th World Filtration Congress*, 70–72 Nice, France.

Falletti, F. 1988. A method of treating polluted oil-in-water emulsions or microemulsions. U.S. Patent Appl. 179,365.

Fulmer, T. D. 1985. Membrane chemical recovery—The next conservation step? America's Textiles Dyeing/Finishing, October.

Gaddis, J. L., H. G. Spencer and D. A. Jernigan. 1989. Caustic recovery and recycling at a textile dyeing and finishing plant. In *Advances in Reverse Osmosis and Ultrafiltration*, Eds. T. Matsuura and S. Sourirajan, pp. 347–355. National Research Council of Canada, Ottawa.

Guibaud, J. 1989. Evaluation of Membralox® ceramic membranes for the regeneration of machining coolants. Societe Ceramiques Techniques, (unpublished work).

Guibaud, J. 1991. (Personal communication).

Guizard, C., J. A. Alary, A. Larbot, L. Cot, M. Rumeau, B. Castelas and J. Gillot. 1984. New mineral membranes for ultrafiltration: application to water purification (in French). *Le Lait.* 64: 276–85.

Guizard, C., A. Larbot and L. Cot. 1990. A new generation of membranes based on organic-inorganic polymers. *Proc. 1st Intl. Conf. Inorganic Membranes*, 3–6 July, 55–64. Montpellier.

Jonsson, A. and E. Petersson. 1988. Treatment of C-stage and E-stage effluents from a bleach plant using a ceramic membrane. *Nordic Pulp and Paper Research Journal* 1: 4–7.

Mir, L., W. Eykamp and R. L. Goldsmith. 1977. Current and developing applications for ultrafiltration. *Ind. Water Engr.* 14(3): 14–19.

Pain, A. T., C. Moulin, M. Faivre, M. M. Bourbigot and M. Rumeau. 1990. Iron removal in groundwater by crossflow micro- and ultrafiltration. *Proc. 5th World Filtration Congress*, 29–35. Nice, France.

Rumeau, M. 1991. (Personal communication).

Soma, C., M. Rumeau and C. Sergent. 1990. Use of mineral membranes in the treatment of textile effluents. *Proc. 1st Intl. Conf. Inorganic Membranes*, 3–6 July, 523–526. Montpellier.

Spencer, H. G. 1990. Formed-in-place inorganic membranes: Properties and applications. *Proc. 1st. Intl. Conf. Inorganic Membranes*, 3–6 July, 95–104. Montpellier.

APPENDIX

INORGANIC MEMBRANE MANUFACTURERS

Organization	Location	Products
Anotec/Alcan	Anotec Separations Ltd. Wildmere road Banbury Oxon, OX16 7JU UK Anotec Separations 226 East 54th Street New York, NY 10022 USA	Ceramic disc filters MF and UF
Asahi Glass	2-1-2 Marunouchi Chiyoda-ku, Tokyo 100 Japan 1185 Avenue of the Americas, 20th Floor New York, NY 10036 USA	Tubular SiO_2-based microfilters
Carbone-Lorraine	Le Carbone Lorraine Tour Manhattan Cedex 21, F-92095 Paris La Defense 2 France Carbone-Lorraine Industries Corporation 400 Myrtle Avenue Boonton, NJ 07005 USA	Tubular MF and UF carbon membranes
Carre/Du Pont	Du Pont Separation Systems; Glasgow Wilmington, DE 19898 USA	Dynamic ceramic and composite membranes on tubular support

Ceramem	Ceramem Corporation 12 Clematis Avenue Waltham, MA 02154 USA	Honeycomb multichannel ceramic membranes on microporous coerdierite support
Ceram Filtre	FITAMM B.P. 60 34402, Lunel Cedex France	Tubular SiC membranes for MF and UF
Du Pont	Du Pont Separation Systems, Glasgow Wilmington, DE 19898 USA	PRD-86 spiral-wound ceramic microfilters
Fairey	Fairey Industrial Ceramics, Ltd. Filleybrooks Stone Staffs ST15 OPU UK	Porous tubular and disc ceramic microfilters
Fuji Filters	Fuji Filter Co., Ltd. 2-4 Nihonbashi- Muromachi, Chuo-ku Tokyo 103 Japan	Porous tubular glass filters for MF and UF
Gaston County	Gaston County Filtration Systems P.O. Box 308 Stanley, NC 28164 USA	Dynamic ZrO_2 membranes on carbon tubes for UF
Hoogovans	Hoogovans Industrial Ceramics BV Postbus 10000 1970 CA Ijmuiden The Netherlands	Tubular ceramic membranes

Imeca/CTI	Imeca B.P. 94 34800 Clermont L'Herault France CTI B.P. 12 36500 Buzancais France	Tubular ceramic membranes for MF
Kubota	Kubota Ltd. 1-2-47 Shikitsu-higashi Naniwa-ku, Osaka 556 Japan	Tubular composite Al_2O_3 membranes for MF and UF
Mott	Mott Metallurgical Corporation Farmington Industrial Park Farmington, CT 06032 USA	Disc and tubular metallic microfilters
NGK	NGK Insulators, Ltd. Shin Maru Building 1-5-1, Marunouchi Chiyodo-ku, Tokyo 100 Japan	Tubular and multichannel ceramic MF and UF membranes
Norton/Millipore	Norton Company 1 New Bond Street Worcester, MA 01606 USA Millipore Corporation Ashby Road, Bedford MA 01730 USA	Ceraflo™ tubular and multichannel ceramic MF and UF membranes
Osmonics	Osmonics, Inc. 5951 Clearwater Drive Minnetonka, MN 55343 USA	Ceratrex™ ceramic microfilters (disc and tubes)
Pall	Pall Porous Metal Filters Courtland, NY 13045 USA	Disc and tubular metallic microfilters

Poral/Pechiney	Voie des collines 38800 Le Pont de Claix France	Metallic and composite ceramic tubes for MF and UF
Schott	Schott Glaswerke Postfach 2480 D-6500 Maniz 1 Germany	Bioran® tubular SiO_2-based ultrafilters
SCT/Alcoa	Societe Ceramiques Techniques, Usine de Bazet B.P. 113, 650001, Tarbes France Alcoa Separations Technology, Inc. 181 Thorn Hill Road Warrendale, PA 15086 USA	Membralox® tubular and multichannel ceramic MF and UF membranes
Tech Sep Rhone Poulenc (formerly SFEC)	Boite Postale No. 201, 84500 Bollene France	Carbosep® ceramic MF and UF tubular membranes
TOTO	Toto Co. Ltd. 1-1 Nakajima 2-Chome Kokura-ku Kita-kyashu-shi 802 Japan	Tubular composite ceramic filters for MF and UF

INDEX